QUANTUM MECHANICS AND THE MRI MACHINE

Quantum Mechanics and the MRI Machine

Stephen G. Odaibo[1,2]
M.S.(Math), M.S.(Comp. Sci.), M.D.

[1]*Quantum Lucid Research Laboratories*
Arlington, VA

[2]*Howard University Hospital*
Washington, D.C.

Copyright ©2012 by Stephen G. Odaibo
All rights reserved. No part of this book may be reproduced or transmitted in any form or by any means, electronic or mechanical, including photocopying, recording, or by any information storage and retreival system, without permission in writing from the publisher.

First published in the United States of America by Symmetry Seed Books.

For information about permissions to reproduce selections from this book, write to Permissions, Symmetry Seed Books, P.O.Box 2173, Arlington VA, U.S.A. 22202

Library of Congress Control Number: 2012951815

ISBN-13: 978-0615708522

ISBN-10: 0615708528

To my Dad

Feedback and Errata
Please kindly email suggestions, errors, and other feedback to QM@qlucid.com

Contents

Contents	viii
List of Tables	x
List of Figures	xi
Acknowledgements	xiii
Preface	xv

1 Introduction   3
    1.1 Quantum Mechanics 4
    1.2 MRI 5
    1.3 Content Outline 5

2 Magnetic Resonance Imaging   7
    2.1 A Hierarchy of Scales 8
    2.2 Magnetism of Electrons and Nucleons 9
    2.3 The Magnetic Potential Energy 11
    2.4 Pulse Magnetization 14
    2.5 Equilibrium Magnetization 15
    2.6 Spin-Lattice Relaxation 16
    2.7 Spin-Spin Relaxation 17
    2.8 Inversion Recovery 18
    2.9 Free Inductance Decay 19
    2.10 Bloch Equations 21
    2.11 Intravenous Contrast Agents 22
    2.12 3D Gradient Localization 24
    2.13 MRI Machinery 25
    2.14 MRI Summary 28

3 Acute Ischemic Stroke   33

3.1 Diffusion-Weighted Imaging 34
3.2 Perfusion-Weighted Imaging 37
3.3 Combined DWI and PWI 37
3.4 Magnetic Resonance Spectroscopy 38
3.5 BOLD MRI 40
3.6 Magnetic Resonance Angiography 42

4 The Hydrogen Atom 45
4.1 Electronic Orbital Configuration 46
4.2 Spherical Harmonics and Radial Waves 59

5 Intrinsic Spin 65
5.1 Spin Algebra 66
5.2 Spin Alignment Geometry 68

6 Clebsch-Gordan Coefficients 71
6.1 Addition Algorithm 72

7 Group Theory: SO(3), SU(2), and SU(3) 75
7.1 SO(3) 76
7.2 SU(2) 77
7.3 SU(2) is Isomorphic to the 3-Sphere 77
7.4 SU(3) 78

8 MRI Signals Processing 81
8.1 Continuous and Discrete Fourier Transforms 82
8.2 MRI Sampling and the Aliasing Problem 83
8.3 The Convolution Theorem 85

9 Future Research Directions 87
9.1 Quantum Information Theory 88
9.2 Quantum Entanglement 89
9.3 Quantum Computing 90
9.4 Quantum Cyber Security 91
9.5 Quantum Optics 92

Bibliography 93

Index 101

List of Tables

2.1 Physical Constants of MRI 30
2.2 Gyromagnetic Ratios of some Biological Nuclei 31

4.1 Spherical Harmonics for $l = 0, 1, 2,$ and 3. 61
4.2 Radial Wave Functions for $n = 0, 1, 2,$ and 3. 62

List of Figures

1.1 T_2 MRI of Brain with Left Occipital Stroke 4

2.1 T_1 FLAIR of Brain 8
2.2 Randomized Spin Orientations 9
2.3 Equilibrium Magnetization 11
2.4 Equilibrium Magnetization Vector 14
2.5 RF Pulse Excitation 15
2.6 Phased Precession 15
2.7 T_1 Recovery Mechanism 16
2.8 T_2 Relaxation Mechanism 17
2.9 T_2 FLAIR of Left PCA and MCA Stroke 18
2.10 Free Inductance Signal 19
2.11 FID Simulation, $\omega = 1$, $T_2 = 1$ 20
2.12 FID Simulation, $\omega = 1$, $T_2 = 2$ 21
2.13 FID Simulation, $\omega = 1$, $T_2 = 4$ 22
2.14 FID Simulation, $\omega = 1$, $T_2 = 8$ 23
2.15 FID Simulation, $\omega = 1$, $T_2 = 16$ 24
2.16 Gradient Coils 25
2.17 X Gradient Coil 26
2.18 Y Gradient Coil 27
2.19 Z Gradient Coil 28
2.20 MRI Machine 29

3.1 ADC Map: Hypointensity in Left MCA Distribution . . 34
3.2 ADC Map: Hypointensity in Left PCA Distribution . . 35
3.3 DWI of Left PCA and MCA Distribution Stroke 36
3.4 2D TOF FSPGR of Carotids 38
3.5 3D TOF of Carotids 39
3.6 MRA of Circle-of-Willis 40
3.7 MRA of Circle-of-Willis: Lateral View 41
3.8 MRA of Neck 42

3.9	SWI of Left MCA Distribution Stroke	43
3.10	SWI of Left PCA and MCA Distribution Stroke	44
4.1	Spherical Harmonic: Y_1^1	51
4.2	Spherical Harmonic: Y_2^1	52
4.3	Spherical Harmonic: Y_3^1	53
4.4	Spherical Harmonic: Y_3^2	54
4.5	Spherical Harmonic: Y_3^3	55
4.6	Spherical Harmonic: Y_4^1	56
4.7	Spherical Harmonic: Y_4^2	57
4.8	Spherical Harmonic: Y_4^3	58
4.9	Spherical Harmonic: Y_4^4	59
4.10	Spherical Harmonic: Y_5^2	60
5.1	Spin Up Geometry	67
5.2	Spin Down Geometry	68
8.1	3D Sinc Function	84
9.1	Bloch Sphere	88

Acknowledgements

The author thanks his friends Henok T. Mebrahtu and Peter Q. Blair for reading this manuscript and providing helpful feedback. He also thanks them for many enjoyable physics conversations over coffee. He thanks Daniel Ennis for making available his MATLAB code for plotting spherical harmonics. He thanks his lovely wife Lisa for a helpful conversation on the clinical use of magnetic resonance spectroscopy in hypoxic-ischemic injury assessment in the neonate. He especially thanks her for her love and support. The author thanks his entire family, and celebrates his father, Dr. Stephen K. Odaibo, on his birthday. He thanks him for being a kind wonderful father and example.

He thanks those who contributed and are contributing to his education, including: Dr. Robert A. Copeland Jr., Dr. Leslie Jones, Dr. Janine Smith-Marshall, Dr. Bilal Khan, Dr. David Katz, Dr. Earl Kidwell, Dr. William Deegan III and Dr. Melissa Kern, Dr. Brian Brooks, Dr. Samuel Belkin, Dr. Ali Ramadan, Dr. Salman J. Yousuf, Dr. Natalie Afshari, Dr. James Lindesay, Dr. Roxanne Springer, Dr. Benjamin Ntatin, Ms. Pam Spencer, Ms. Asala, Mr. Badamasi, Dr. Xiaobai Sun, Dr. Mark Dewhirst, Dr. Carlo Tomasi, Dr. John Kirkpatrick, Dr. Nicole Larrier, Dr. Brenda Armstrong, Maureen D. Cullins, Ms. Sharon Coward, Mr. Richard Wallace, Dr. Delbert R. Wigfall, Dr. Philip C. Goodman, Dr. Joanne A. P. Wilson and Dr. Ken Wilson, Dr. Srinivasan Mukunduan Jr., Dr. Augustus Grant, Dr. David L. Simel, Dr. Danny O. Jacobs, Dr. Timothy Elston, Dr. Onyiye Akwari and Dr. Anne Akwari, Dr. Barton F. Haynes and Dr. Caroline Haynes, Dr. Cedric Bright, Dr. Ayodeji Ayo-Bello and Dr Yetunde Ayo-Bello, Dr. Mark Branch, Dr. Larry Greenblatt, Dr. Leon Herndon, Dr. Terri Young, Dr. Alan Carlson, Dr. Diana McNeil, Dr. Aimee Zaas, Dr. Kenneth Goldberg, Dr. Delano Meriwether and Novimbi Meriwether, Dr. Jean Spaulding, Dr. Charles Johnson, Dr. David

F. Bell, Dr. Jacqueline Looney, Dr. Calvin R. Howell, Dr. Roxanne Springer, Dr. Salvatore Pizzo, Dr. Dona Chikaraishi, Dr. Necmettin Yildrim, Dr. Eric Reiter, Dr. Scott Deken, Dr. John Mayer, Dr. Lex Oversteegen, Dr. Peter V. O'Neil, Dr. Anne Cusic, Dr. Henry van den Bedem, Robin Geurs, Christina Banks, Stacye Fraser Thompson, Dennis Aggrey, Dr. Rudi Weikard, Dr. James R. Ward, Dr. Marius Nkashama, Dr. Michael W. Quick, Dr. Robert J. Lefkowitz, Dr. Arlie O. Petters, and several other excellent educators, mentors, role models, friends, and family not mentioned here.

Preface

In this book, I explain Quantum mechanics, Magnetic resonance imaging (MRI), and how the later arises from the former. I carefully traverse the hierarchy of scales from the spin and orbital angular momentum of subatomic particles to the ensemble magnetization of tissue. A number of modalities used in the assessment of acute ischemic stroke are reviewed along the way. The book's step-by-step approach makes it well suited to undergraduate and begining graduate-level learners of Physics, as well as to anyone interested in acquiring a fundamental understanding of the mechanism of magnetic resonance technology. For instance, researchers in various areas of Biomedical engineering, Electrical engineering, Computer science, NMR Chemistry, Radiology, Radiation oncology, and Neurology, will all find this a useful resource. Familiarity with Linear algebra and Differential equations is assumed.

I emphasize the central relevance of the Hydrogen atom to MRI. And I present the analytical solution of the Hydrogen Schrödinger equation in a step-by-step way that deliberately avoids the cryptographic conciseness of several Quantum mechanics textbooks which often yields confusion, frustration, and angst in learners. Skipping no steps, I explain how the Spherical harmonics and Radial wave functions arise, and what they mean. In the explanation of magnetic resonance, each step in the mechanism is accompanied by a picture illustration. The final chapter provides a glimpse of a certain exciting side of the applied physics frontier. That chapter briefly overviews quantum information, quantum computing, quantum cryptography, quantum entanglement, and quantum optics. This initial edition of the book introduces key concepts, but does not contain excercises or worked examples. As with any subject, mastery requires active problem solving and lots of practice; and as such this edition is best used as a supplement to traditional textbooks.

This book is inherently interdisciplinary, and therefore the savvy reader is welcome to skip around where necessary. For instance, some graduate or advanced undergraduate learners of Physics may find the step-by-step solution of the Hydrogen Schrödinger equation verbose. On the other hand, they may appreciate the gentleness of the explanations on MRI modalities used in the assessment of acute ischemic stroke. A Neuroradiologist or a Biomedical engineering researcher on the other hand may have the exact opposite experience.

The book is interluded with the ancient whispers of a great scientist, Michael Faraday; arguably the most meticulous experimentalist ever, and a deeply insightful natural philosopher indeed. Though Quantum mechanics was formulated 40 to 50 years after Faraday's death, his early insights and musings in electricity and magnetism are just as relevant and perhaps even more instructive today. Almost two centuries after his discoveries of electromagnetic induction and the magnetism of light, both are at work in MRI machines, saving lives everyday around the world.

—Stephen G. Odaibo, M.D.

Chapter 1

Introduction

Electricity is often called wonderful, beautiful; but it is so only in common with the other forces of nature. The beauty of electricity or of any other force is not that the power is mysterious, and unexpected, touching every sense at unawares in turn, but that it is under law, and that the taught intellect can even govern it largely. The human mind is placed above, and not beneath it, and it is in such a point of view that the mental education afforded by science is rendered super-eminent in dignity, in practical application and utility; for by enabling the mind to apply the natural power through law, it conveys the gifts of God to man.*

—Michael Faraday

---

*Lecture notes from an 1858 Friday discourse at the Royal Institution, from *The Life and Letters of Faraday* (1870), H. B. Jones (ed.), Vol. 2, p. 404

4   CHAPTER 1. INTRODUCTION

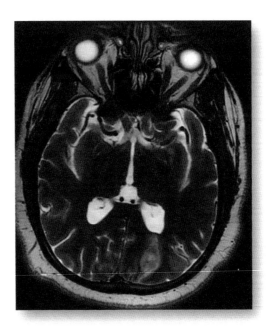

Figure 1.1: T_2 axial fast spin echo image showing acute ischemia in the PCA distribution. There is relative hyperintensity in the left occipital region of infarction. Note the prominent hyperintensity of the vitreous and cerebrospinal fluid, both consistent with the hyperintensity of water on T_2-weighted imaging.

1.1 Quantum Mechanics

Quantum mechanics is a probabilistic physical model for describing subatomic phenomena [34, 43, 101]. It began around 1894 with Max Planck's attempts to understand *Black-body radiation*, and to circumvent the *ultraviolet catastrophe*, a manifest failing of the classical physics model. During the ensuing 30 years, quantum mechanics was gradually formulated by various researchers including Werner Heisenberg, Erwin Schrödinger, Niels Bohr, John von Neumann, David Hilbert, and P.A.M. Dirac. The theory has had great success as a model of electronic atomic orbital configurations, and thereby underlies much of the current framework for understanding of our physical world. Furthermore, quantum mechanics has provided theoretical impetus and insight for a number of clinical applications such as the MRI.

1.2 MRI

MRI plays a key role in the diagnosis and management of acute ischemic stroke and traumatic brain injury. A rigorous understanding of MRI begins with the elementary particles which give rise to the magnetic properties of tissue. Elementary particles have two independent properties which manifest magnetic moments: (i) orbital angular momentum, and (ii) intrinsic spin. These two properties are independent only to first order, as it was their very interaction via spin-orbit coupling which led to detectable energy differences in the hydrogen spectrum, and consequently to the discovery of intrinsic spin. We discuss both properties in this book in the context of the hydrogen atom. The hydrogen atom ^{1}H is the most frequently targeted nucleus in MRI due largely to its biological abundance and high gyromagnetic ratio. The ^{1}H nucleus consists of a single proton. Protons are made of quarks, specifically one down and two up quarks, each of which are of spin 1/2. The charge and spin of the proton are both directly due to its quark composition. The down quark has a charge of -1/3 while the up quarks each have charge +2/3, which all sum up to a charge of +1. The nuclear spin is 1/2 by virtue of spin cancellation from the antiparallel alignment of two of the three quarks.

1.3 Content Outline

The remainder of this book is organized as follows: Chapter (2) presents an overview of the MRI, Chapter (3) reviews the MRI assessment of acute ischemic stroke, Chapter (4) describes the hydrogen atom and reviews an analytical solution to its Schrödinger equation, Chapter (5) reviews intrinsic spin, Chapter (6) reviews quantum mechanical addition of orbital angular momentum and spin, Chapter (7) reviews Group Theory of SU(2) and SO(3) and their relationship, Chapter (8) reviews the basics of MRI signals processing, and Chapter (9) briefly overviews some active areas of quantum mechanics research, and their potential future applications. Specifically, quantum computing, quantum information, and quantum optics.

CHAPTER 1. INTRODUCTION

The MRI Chapter, Chapter (2), contains a number of sections which are outlined as follows: Section (2.1) discusses the hierarchy of scales in MRI phenomenology, Section (2.2) reviews the magnetism of electrons and nucleons, Section (2.3) reviews the magnetic potential energy and its role in the descriptions of magnetization, torque, force, and Larmor precession, Section (2.4) reviews pulse magnetization, Section (2.5) describes equilibrium magnetization, Section (2.6) describes spin-lattice or T_1 recovery, Section (2.7) describes spin-spin or T_2 decay, Section (2.9) reviews free inductance decay, Section (2.10) reviews the Bloch equations, Section (2.11) overviews intravenous contrast agents in MRI, Section (2.12) reviews gradient-based band-selection method for MRI space localization, and Section (2.13) reviews the components of the MRI machine.

The stroke chapter, Chapter (3), contains a number of sections which are outlined as follows: Section (3.1) reviews diffusion weighted imaging (DWI), Section (3.2) reviews perfusion weighted imaging (PWI), Section (3.3) reviews the combination of DWI and PWI and its utility in imaging the penumbra, Section (3.4) reviews magnetic resonance spectroscopy, Section (3.5) reviews blood oxygen level dependent (BOLD) imaging, and Section (3.6) reviews magnetic resonance angiography (MRA).

Chapter 2

Magnetic Resonance Imaging

I have been driven to assume for some time, especially in relation to the gases, a sort of conducting power for magnetism. Mere space is Zero. One substance being made to occupy a given portion of space will cause more lines of force to pass through that space than before, and another substance will cause less to pass. The former I now call Paramagnetic & the latter are the diamagnetic. The former need not of necessity assume a polarity of particles such as iron has when magnetic and the latter do not assume any such polarity either direct or reverse. I do not say More to you just now because my own thoughts are only in the act of formation but this I may say that the atmosphere has an extraordinary magnetic constitution & I hope & expect to find in it the cause of the annual & diurnal variations, but keep this to yourself until I have time to see what harvest will spring from my growing ideas. *

—Michael Faraday

---

*Letter to William Whewell, 22 Aug 1850. In Frank A. J. L. James (ed.), *The Correspondence of Michael Faraday* (1996), Vol. 4, p. 177.

CHAPTER 2. MAGNETIC RESONANCE IMAGING

The MRI is an optical technology whose core equation is the Faraday-Maxwell Equation. It exploits ensemble phenomena in which the composition of a sample can be probed by sensing its magnetic properties through radio frequency (RF) waves. Here, we describe the various parts of MRI and their respective roles in the whole.

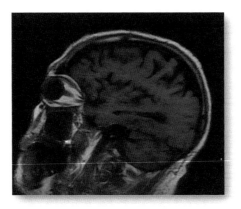

Figure 2.1: T_1 FLAIR. Sagittal section. Note the hyperintensity of orbital fat, and the hypointensity of vitreous and cerebrospinal fluid. Hyperintensity of fat and hypointensity of water are radiologic characteristics of T_1 weighted imaging.

2.1 A Hierarchy of Scales

The hierarchy of scales in the modeling of magnetic resonance imaging proceeds from the subnuclear scale of quarks and gluons, to the subatomic scale of discrete electrons and nucleons, and finally to the bulk sample scale where ensemble effects and statistical mechanics are the operative physics. In a general sense, the subnuclear scale is coupled to quantum field theory, the subatomic scale is coupled to quantum mechanics, and the bulk sample scale is coupled to classical electromagnetics.

Certain phenomena in MRI are only meaningful over specific regimes. For instance, a magnetic moment experiences a torque when an external magnetic field is applied. However, the concept of torque is a classical concept. It is deterministic and continuous, and is meaningful only at the level of ensemble or bulk effect. On the subatomic level, such as on the scale of single electrons and nucleons, quantum mechanics is the operative physics, and state transitions are quantized. This is encoded in the time evolution of quantum states governed by an appropriate Hamiltonian, yielding the specific Schrödinger equation that applies in that setting.

Although pedagogically separable, the scales are intrinsically linked physically. For instance, the bulk magnetization, **M**, is the net sum of the atomic level magnetic moments, $\boldsymbol{\mu}_j$. Similarly, the pulse magnetization frequency used to torque the bulk magnetization vector and change its direction is a radio frequency wave, **B**$_1$, oscillating at the Larmor frequency of the individual nucleons. Furthermore, **B**$_1$ takes its effect by acting directly on the individual electrons and nucleons, causing them to precess in phase.

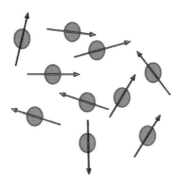

Figure 2.2: Randomized spin orientations. Nuclear spin orientations are random in the absence of an external magnetic field.

2.2 Magnetism of Electrons and Nucleons

Electrons, protons, and neutrons are magnets. They derive their magnetism from intrinsic spin and orbital angular momentum. In the absence of an external magnetic field, each particle's spin is oriented in an essentially arbitrary direction, as illustrated in Figure (2.2). However, in the presence of an externally applied magnetic field, a bulk magnetism of sample tissue occurs via spin alignment, as illustrated in Figure (2.3). Both Figures are 2-dimensional projections of the 3-dimensional spin direction vectors. The magnetic dipole moment, $\boldsymbol{\mu}$, of a single unpaired electron or nucleon is given by,

$$\boldsymbol{\mu} = \gamma \mathbf{S}, \tag{2.1}$$

where γ is the gyromagnetic ratio measured in Hz/T (Hertz per Tesla) and $\mathbf{S}$ is the spin observable. The spin observable has the following two defining properties,

$$\mathbf{S}^2 |s, m\rangle = \hbar^2 s(s+1) |s, m\rangle, \tag{2.2}$$

and

$$\mathbf{S}_z |s, m\rangle = \hbar m |s, m\rangle, \tag{2.3}$$

where $|s, m\rangle$ is a spin eigenstate vector for a particle of spin s and magnetic spin quantum number m, $\mathbf{S}^2$ is the square of the spin, and $\mathbf{S}_z$ is the z-component of spin. In general, $m \in \{-s, -(s-1), ..., s-1, s\}$, and in particular for the electron and proton $s = 1/2$ and $m \in \{-1/2, 1/2\}$. Spin is further discussed in Chapter (5). The gyromagnetic ratio for the electron is given by,

$$\gamma_e = g_e \frac{e}{2m_e}, \tag{2.4}$$

where e is the elementary charge, m_e is the mass of the electron, and g_e is the electron gyromagnetic factor (g-factor), a dimensionless number whose experimental agreement with theory is near unprecedented on this scale, and thereby lends strong credibility to the theory of quantum electrodynamics. Table (2.1) shows the electron and proton g-factors. The gyromagnetic ratio for nucleons is given by,

$$\gamma_n = g_p \frac{e}{2m_p}, \tag{2.5}$$

where m_p is the mass of the proton, and g_n is the g-factor of the nucleon. Estimated gyromagnetic ratios of some biologically-relevant nuclei are shown in Table (2.2) [14, 64].

The Bohr magneton, μ_B, and the nuclear magneton, μ_N, are named quantities related to the electron and proton gyromagnetic ratios respectively, and are given by,

$$\mu_B = \frac{e\hbar}{2m_e}, \tag{2.6}$$

2.3. THE MAGNETIC POTENTIAL ENERGY

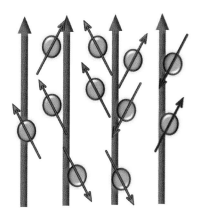

Figure 2.3: Nuclear spins align parallel or antiparallel to an applied external magnetic field.

and

$$\mu_N = \frac{e\hbar}{2m_p}. \tag{2.7}$$

The magnetic moment of the electron is much larger than that of the nucleons, and as is discussed below, this corresponds to smaller precession, pulse, and signal frequency for protons than electrons. Specifically, nucleons precess, absorb, and emit electromagnetic signals in the radio frequency range. This range is non-ionizing radiation, and makes MRI a safer choice than ionizing imaging modalities such as plane and computed tomography X-rays.

2.3 The Magnetic Potential Energy

The magnetic potential energy is,

$$U = -\boldsymbol{\mu} \cdot \mathbf{B}. \tag{2.8}$$

The force on a magnetic moment, the torque on a magnetic moment, and the Larmor precession frequency or resonance frequency, can all be described in terms of the magnetic potential energy, $U(\mathbf{r}, t)$.

Magnetization

The magnetic moment aligns itself in either parallel or anti-parallel orientation to the external field $\mathbf{B}_0$. This process is called magnetization. In the case of a population sample, upon application of $\mathbf{B}_0$, the spin orientations go from a somewhat arbitrary array of directions to being composed of only two directions, parallel and antiparallel. See Figure (2.4). The parallel orientation is the lower energy state, and thereby is the more populous state for protons in the sample. The population distribution is temperature dependent, such that the proportion of protons in the higher energy state increases with temperature. The relationship is given by,

$$\frac{N_h}{N_l} = e^{-E/kT}, \qquad (2.9)$$

where N_h is the higher energy state, N_l is the lower energy state, E is the energy difference between the two states, k is the Boltzmann constant, and T is the absolute temperature in Kelvin. The bulk magnetization can be modeled as,

$$\mathbf{M}(t) = \sum_j \boldsymbol{\mu}_j(t). \qquad (2.10)$$

with $\mathbf{B}_0 = \hat{\mathbf{z}} B_0$, it follows that $M(0) = M_z(0)$ and $M_{xy} = 0$. In other words, the equilibrium magnetization is entirely along the z (longitudinal) direction and there is no component in the xy (transverse) direction.

Torque

The torque on a magnetic moment in a magnetic field acts to align it with the field, is a function of the magnetic potential energy, and is given by,

$$\begin{aligned}
\tau(\alpha) &= -\frac{\partial}{\partial \alpha}(-\boldsymbol{\mu} \cdot \mathbf{B}) \\
&= -\frac{\partial}{\partial \alpha}(-\mu B \cos \alpha) \\
&= \mu B \frac{\partial \cos \alpha}{\partial \alpha} \\
&= -\mu B \sin \alpha \\
&= -|\boldsymbol{\mu} \times \mathbf{B}|,
\end{aligned} \qquad (2.11)$$

2.3. THE MAGNETIC POTENTIAL ENERGY

where α is the angle between $\mathbf{B}$ and $\boldsymbol{\mu}$, torque direction is according to the right hand rule, and the minus sign indicates the restoring nature of the torque.

Force

The force on a nucleon or electron in a magnetic field is,

$$\mathbf{F} = -\nabla(-\boldsymbol{\mu} \cdot \mathbf{B}). \quad (2.12)$$

Precession

When an external magnetic field $\mathbf{B}_0$ is applied to an electron or neutron particle, the particle's magnetic moment precesses about the direction of $\mathbf{B}_0$ at a frequency called the Larmor frequency given by,

$$\omega = \gamma B_0. \quad (2.13)$$

In the equilibrium magnetization state with $\mathbf{B}_0 = \hat{\mathbf{z}} B_0$ and $M_{xy} = 0$, there is no precession of the bulk magnetization $\mathbf{M}$ because $\mathbf{M} \times \mathbf{B}_0 = 0$. Figure (2.4) illustrates the magnetization of the individual spins, their precession about the applied field direction, and the bulk magnetization vector.

Considering only the spin of a spin 1/2 object in a magnetic field, and assuming $\mathbf{B} = B\hat{\mathbf{e}}_z$, we get the following Hamiltonian,

$$H = -\boldsymbol{\mu} \cdot \mathbf{B} = -(\frac{-g_x e}{2m_x}\mathbf{S}) \cdot \mathbf{B} := \boldsymbol{\omega} \cdot \mathbf{S}, \quad (2.14)$$

where the subscript x denotes either e for electron or p for a neutron, and where,

$$\boldsymbol{\omega} := \frac{g_x e B}{2m_e}\hat{\mathbf{e}}_z. \quad (2.15)$$

It follows that,

$$H = \omega S_z. \quad (2.16)$$

We see that remarkably, the Hamiltonian is proportional to the z-component of spin, therefore the magnetic spin eigenstates are simultaneously energy eigenstates. Furthermore the energy eigenvalues are proportional to the magnetic spin quantum numbers. Specifically,

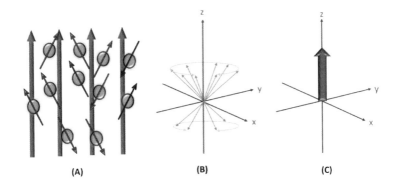

Figure 2.4: Figs (A) and (B) show individual nuclei aligned parallel or antiparallel to the applied external magnetic field. Note that more nuclei align parallel because this is the lower energy state. Fig (C) shows the magnetization vector having only longitudinal component. Due to cancellation by dephased oscillation, no transverse component manifests at equilibrium.

$$E_\pm = \pm \frac{1}{2}\hbar\omega. \qquad (2.17)$$

The above relation led early researchers to the discovery of spin, by its manifestation as the Zeeman effect and the splitting of energy levels in the hydrogen optical spectrum.

2.4 Pulse Magnetization

Given a bulk magnetization $\mathbf{M}$ due to a static magnetic field $\mathbf{B}_0 = B_0\hat{\mathbf{z}}$, a pulse of a weaker magnetic field, $\mathbf{B}_1$, oscillating at the Larmor frequency and applied perpendicular to $\mathbf{B}_0$, will result in a transition of the individual nucleons from dephased to in-phase oscillation. See Figure (2.5). This manifests as the acquisition of a transverse phase $\mathbf{M}_{xy}$ and oscillations of $\mathbf{M}$ about the z-axis at the Larmor frequency ω. This tilting of the magnetization vector also implies a decrease in M_z. Both effects are illustrated in Figure (2.6). During the pulse, the net external magnetic field is the sum of $\mathbf{B}_0$ and $\mathbf{B}_1$. If the pulse duration was extended indefinitely, the result would be alignment of the bulk magnetization with the new direction, $\mathbf{B}_0 + \mathbf{B}_1$, and there would be no resultant oscilla-

2.5. EQUILIBRIUM MAGNETIZATION

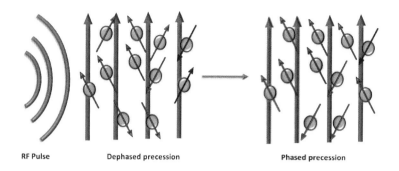

Figure 2.5: RF Pulse excitation. The electromagnetic wave pulse shown on the left, induces a state change from dephased to in-phase oscillation.

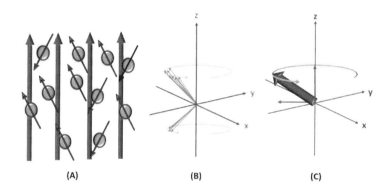

Figure 2.6: Instantaneously after pulse excitation, nuclei begin to oscillate in phase. The magnetization vector acquires a transverse component and also begins to oscillate at the Larmor precession frequency.

tion observed. The pulsed nature of $\mathbf{B}_1$ is therefore critical to the technology of magnetic resonance imaging.

2.5 Equilibrium Magnetization

The net magnetization is given by,

16 CHAPTER 2. MAGNETIC RESONANCE IMAGING

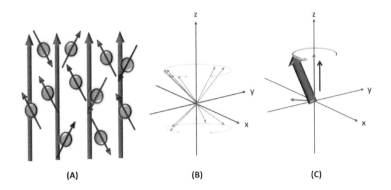

Figure 2.7: Schematic of T_1 Recovery. Figs (A) and (B) show individual nuclei with two up spins and two down spins dephased. Fig(C) shows attenuation of precession of the magnetization vector. The red arrow indicates progressive recovery of longitudinal magnetization. This state is intermediate between dephased and in-phase oscillation states depicted in Figs (2.4) and (2.6) respectively.

$$\mathbf{M} = (N_l - N_h)\boldsymbol{\mu}, \tag{2.18}$$

while the ratio of N_l to N_h is given in Equation (2.9) above as, $\frac{N_h}{N_l} = e^{-E/kT}$. At equilibrium, the solution of the above system of two equations yields the equilibrium magnetization,

$$M_0 = \frac{B_0 \gamma^2 \hbar^2}{4 \mathrm{k} T} \rho_p, \tag{2.19}$$

where $\mathrm{k} = k/2\pi$ is the reduced Boltzmann constant and ρ_p is the proton density.

2.6 Spin-Lattice Relaxation

The spin-lattice relaxation, also known as T_1 relaxation or longitudinal relaxation, is the process of longitudinal magnetization recovery following a $\mathbf{B}_1$ perturbation. Specifically, T_1 is the time constant of longitudinal magnetization recovery, and is given by,

$$M_z(t) = M_0(1 - e^{-t/T_1}) + M_z(0)e^{-t/T_1}, \tag{2.20}$$

2.7. SPIN-SPIN RELAXATION

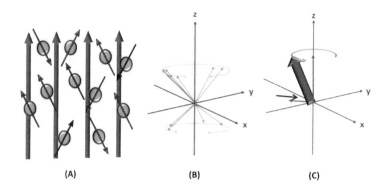

Figure 2.8: Schematic of T_2 Relaxation. Figs (A) and (B) show individual nuclei with two up spins and two down spins dephased. Fig(C) shows attenuation of precession of the magnetization vector. The red arrow indicates progressive loss of transverse magnetization. This state is intermediate between dephased and in-phase oscillation states depicted in Figs (2.4) and (2.6) respectively.

where M_0 is equilibrium magnetization and $M_z(0)$ is the longitudinal magnetization instantaneously after pulse excitation. Figure (2.7) illustrates the T_1 recovery process.

2.7 Spin-Spin Relaxation

The spin-spin relaxation, also known as T_2 relaxation or transverse relaxation, is the process of transverse magnetization relaxation following a $\mathbf{B}_1$ perturbation. Specifically, T_2 is the time constant of transverse magnetization recovery, and is given by,

$$M_{xy}(t) = M_{xy}(0)e^{-t/T_2}, \qquad (2.21)$$

where $M_{xy}(0)$ is the transverse magnetization instantaneously after pulse excitation. Following the B_1 induction of in-phase precession, the proton spins again begin to dephase according to T_2 time constant. Figure (2.8) illustrates the T_2 relaxation process. Of note, the relaxation of transverse magnetization is faster than the recovery of longitudinal magnetization. In other terms, $T_2 < T_1$. Figure (1.1) shows a T_2-weighted fast spin echo image of acute ischemic stroke in the posterior circulation.

18   CHAPTER 2. MAGNETIC RESONANCE IMAGING

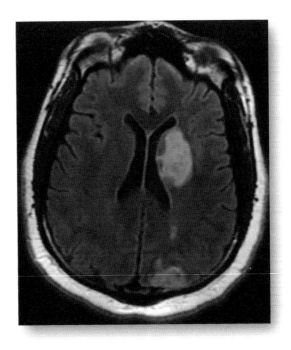

Figure 2.9: T_2 FLAIR showing hyperintense signal in the PCA and MCA distributions, consistent with acute ischemic stroke. Note that the hypointensity of cerebrospinal fluid due to fluid-attenuation allows for enhanced contrast with the periventricular lesion.

2.8 Inversion Recovery

In inversion recovery, the initial magnetization vector is negative, and typically $M_z(0) = -M_0$. Substituting this value into Equation (2.20) above yields,

$$M_z(t) = M_0(1 - 2e^{-t/T_1}). \qquad (2.22)$$

Fluid attenuated inversion recovery (FLAIR) is useful in the diagnosis of certain pathologies such as periventricular white-matter lesions in multiple sclerosis. FLAIR can be T_1 or T_2 weighted. Figure (2.1) shows a sagittal section of a T_1 FLAIR brain image,

2.9. FREE INDUCTANCE DECAY

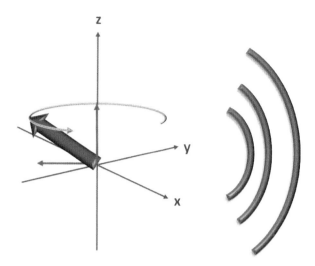

Figure 2.10: Oscillation of the magnetization vector gives rise to an electromotive force which drives a current in the RF coils.

and Figure (2.9) shows a T_2 FLAIR brain image.

2.9 Free Inductance Decay

Upon pulse excitation and tilting, the magnetization vector oscillates about the z-axis. This oscillating magnetic moment in turn generates an electromotive force in accordance with the Faraday-Maxwell law of electromagnetic induction,

$$\nabla \times \mathbf{E} = -\frac{\partial \mathbf{B}}{\partial t}, \tag{2.23}$$

where $\nabla \times$ is the curl operator, $\mathbf{E}(\mathbf{r}, t)$ is the electric field, and $\mathbf{B}(\mathbf{r}, t)$ is the magnetic field. See Figure (2.10). The above equation can be rendered in integral form via Stokes theorem to yield,

$$\oint_{\partial \Sigma} \mathbf{E} \cdot d\mathbf{s} = -\frac{\partial}{\partial t} \iint_{\Sigma} \mathbf{B} \cdot d\mathbf{A}, \tag{2.24}$$

where $\partial \Sigma$ can represent a wire coil bounding a surface Σ, $d\mathbf{s}$ is an infinitesimal segment length along the wire, and $d\mathbf{A}$ is vector normal to an infinitesimal area of Σ. Such integral transformation

20 CHAPTER 2. MAGNETIC RESONANCE IMAGING

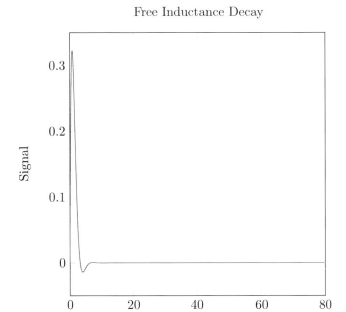

Figure 2.11: FID signal simulated with a 1 rad/sec sinusoid, $T_2 = 1$, and $M_{xy}(0) = 1$ secs.

has certain advantages in numerical computation. For instance, Green's functions approaches can be invoked and can greatly decrease the number of unknowns, thereby decreasing the computational resource demand of the problem.

The RF coils within the MRI probe experience an electromotive force due to the rotating magnetization. This results in an electric current signal which is fed as output to a screen. The signal decays with time according to the T_2 time constant, hence the name Free Inductance Decay (FID) signal [105]. Figures (2.11) to (2.13) are simulations of FID signals, showing the effect of progressively greater T_2 time constant. Specifically, the T_2 values are varied from 1 to 16 secs, while the precession frequency ω is held constant at 1 radians/sec. Though the chosen simulation parameters differ significantly from ranges typical of biological systems, they do highlight both the characteristic shape and T_2 dependence of the waveform and envelope. Physiological systems often have

2.10. BLOCH EQUATIONS

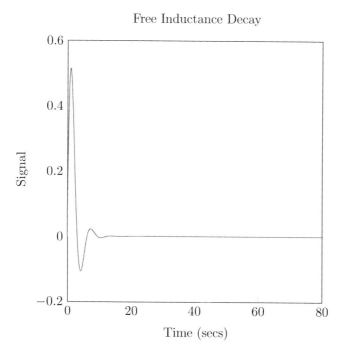

Figure 2.12: FID signal simulated with a 1 rad/sec sinusoid, $M_{xy}(0) = 1$, and $T_2 = 2$ secs.

parameters in the range: $10^7 \leq \omega \leq 3 \times 10^8$ radians/sec, and $10^{-2} \leq T_2 \leq 3$ secs.

2.10 Bloch Equations

The Bloch equations are a phenomenological description of the time evolution of bulk magnetization, and are also referred to as the equations of motion of magnetization. They are given by,

$$\frac{\partial M_x(t)}{\partial t} = \gamma (\mathbf{M}(t) \times \mathbf{B}(t))_x - \frac{M_x(t)}{T_2}, \quad (2.25)$$

$$\frac{\partial M_y(t)}{\partial t} = \gamma (\mathbf{M}(t) \times \mathbf{B}(t))_y - \frac{M_y(t)}{T_2}, \quad (2.26)$$

and
$$\frac{\partial M_z(t)}{\partial t} = \gamma (\mathbf{M}(t) \times \mathbf{B}(t))_z - \frac{M_z(t) - M_0}{T_1}, \quad (2.27)$$

22 CHAPTER 2. MAGNETIC RESONANCE IMAGING

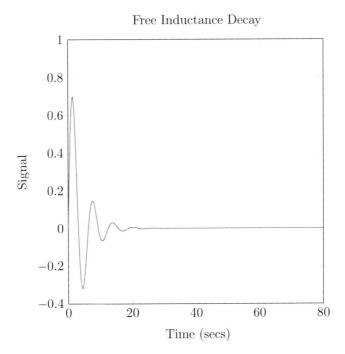

Figure 2.13: FID signal simulated with a 1 rad/sec sinusoid, $M_{xy}(0) = 1$, and $T_2 = 4$ secs.

where γ is the gyromagnetic ratio of the nucleon and $\mathbf{B}(t) = \mathbf{B}_0 + \mathbf{B}_1(t)$.

2.11 Intravenous Contrast Agents

There are several intravenous MR contrast agents in current clinical use, most of which are based either on low molecular weight chelates of gadolinium ion (Gd^{3+}) [24, 17] such as Gadolinium-DPTA, or are based on iron oxide (FeO) such as Fe_3O_4, Fe_2O_3, and γ-Fe_2O_3 (gamma phase maghemite) [18, 5, 62, 53]. By virtue of their magnetic properties, they confer contrast of vasculature from background, and by proxy of vascularization density, of organs from surround. Specifically, the Gd^{3+}-based agents predominantly shorten T_1 time, while the FeO-based agents predominantly shorten T_2 time. Nanoparticle formulations of contrast agents are in various stages of development and can be based on gadolinium

2.11. INTRAVENOUS CONTRAST AGENTS

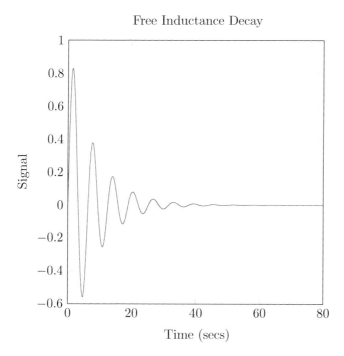

Figure 2.14: FID signal simulated with a 1 rad/sec sinusoid, $M_{xy}(0) = 1$, and $T_2 = 8$ secs.

or iron-oxide, or on other agents such as cobalt, nickel, manganese, or copper ions [113, 60, 5]. Optimization of the efficacy of MR contrast agents is an active area of research [22, 54, 57, 96, 77, 52, 58]. The growing array of MR contrast applications has included efforts to target tissue or tumor-specific receptors [58, 65, 23, 117, 39, 82, 107]. Serious adverse reactions to intravenous MRI contrast agents are rare. However in the mid 2000s gadolinium chelates were increasingly hypothesized as playing a role in nephrogenic systemic fibrosis (NSF), a disease entity unknown prior to 1997. It has since been essentially established that Gadodiamide, a gadolinium-containing agent, is associated with an increased risk of developing NSF in susceptible individuals [13, 59, 100, 110, 72]. Superparamagnetic iron oxides have been promoted by some as possible alternatives to gadolinium-based agents in patients at risk for NSF [86].

24   CHAPTER 2.   MAGNETIC RESONANCE IMAGING

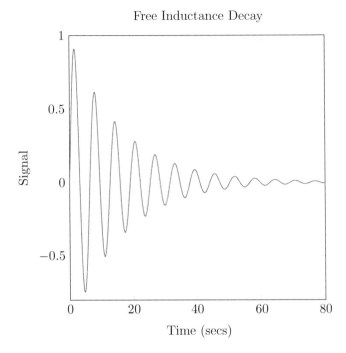

Figure 2.15: FID signal simulated with a 1 rad/sec sinusoid, $M_{xy}(0) = 1$, and $T_2 = 16$ secs.

2.12   3D Gradient Localization

For 3D localization, a gradient coil generates a magnetic field gradient in the bore. Then in accordance with the Larmor frequency relation $\omega = \gamma B(r, t)$, it follows that the resonance frequency, $\omega = \omega(r, t)$ acquires space dependence, and thereby serves to effectively label the tissue in space. The process of space-specific excitation of the sample is called *slice selection*. The excitation pulse frequency is simply chosen as a bandwidth frequency flanking the region-of-interest or the *slice*. The process of *scanning* is an iteration, however simple or complex, through a sequence of slices. And slice thickness is proportional to excitation pulse bandwidth. The slice thickness can be decreased by decreasing the bandwidth, or alternatively by increasing the magnetic field gradient. For instance, given a constant magnetic field gradient, the slice thickness between two points r_1 and r_2 is given by,

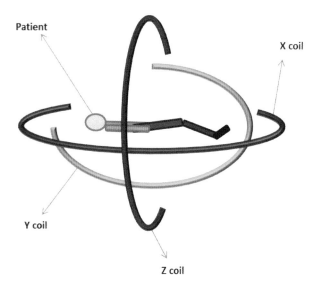

Figure 2.16: Gradient coils: X, Y, and Z.

$$\frac{\omega(r_2) - \omega(r_1)}{\gamma \nabla B} = r_2 - r_1 \qquad (2.28)$$

Figure (2.16) is a schematic of the gradient coils, and Figures (2.17) to (2.19) show the three directional component coils. Both the bandwidth and magnetic field gradient are Graphical User Interface (GUI)- adjustable parameters on current day MRI machines. The gradient-based localization technique was Paul Lauterbur's contribution to the development of MRI [61].

2.13 MRI Machinery

The MRI is a simple machine in principle. It was invented by Raymond V. Damadian by 1971 [31, 33, 32, 115]. It consists of a magnet, a radio frequency coil, an empty space or bore, and a computer processor. A schematic illustration of the MRI machine is shown in Figure (2.20). The magnet is for applying the static and pulsed magnetic fields. Current day machines typically use superconducting electromagnets. These are wire coils cooled to low temperatures using liquid helium and liquid nitro-

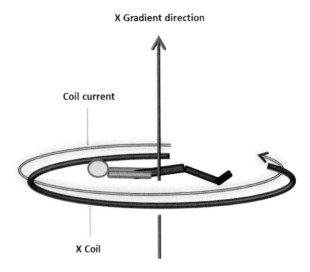

Figure 2.17: X Gradient coil. Coil is perpendicular to gradient direction in accordance with Faraday-Maxwell Equation.

gen [30, 67, 15, 51, 48, 111, 76, 115], allowing for greatly decreased electrical resistance, and consequently high current loops generating strong magnetic fields. The RF coil typically serves both as the generator of the pulsed magnetic field, and as the detector of the FID signal. Such duality of function is possible simply because the Faraday-Maxwell equation is true in both directions. Several machines also have a separate gradient coil. The gradient coil has x, y, and z components indicating the spatial direction of the gradient field, and together conferring 3D localization. The computer processor runs the software instructions for pulse sequences and signals processing, including image reconstruction.

Superconductivity

MRI requires strong magnetic fields. The Earth's magnetic field is variable with time, is asymmetric between northern and southern hemispheres [89], and is estimated to range between 25,000 and 65,000 nanoteslas (0.25 to 0.65 Gauss) at the surface, with a commonly cited average of 46,000 nanotesla (0.46 Gauss) [118]. Current clinical-use MRIs commonly range from 1.5 Tesla to 7

2.13. MRI MACHINERY

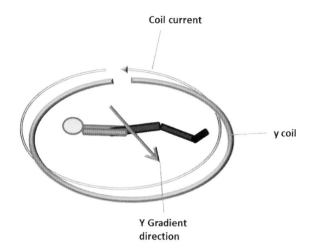

Figure 2.18: Y Gradient coil. Coil is perpendicular to gradient direction in accordance with Faraday-Maxwell Equation.

Teslas. Therefore MRIs may require magnetic fields up to 280,000 times the earth's magnetic field. This has been achieved by a shift from permanent magnets to superconducting electromagnets, and presents another interdisciplinary link to both theory and practice. The dominant physical theory of superconductivity is called BCS theory [8, 26, 9, 10] after its formulators, Bardeen, Cooper, and Schrieffer, and is based on electron pairing via phonon-exchange to form so called Cooper pairs. Though electrons are fermions and subject to the Fermi exclusion principle, precluding any two electrons occupying the same state, Cooper pairs act like bosons because their pairing and spin summation result in integer-spin character, a type of boson-like state. Upon cooling, these boson-like pairs can condense, analogous to Bose-Einstein condensation and representing a quantum phase transition. Quantum phase transitions have historically been challenging to study in a controlled way in the laboratory, but have recently been demonstrated by H. T. Mebrahtu et. al. in tunable quantum tunnelling studies of Luttinger liquids [75]. In superconductivity, upon transition into the condensed state, the coherence of the system imposes a high energy penalty on resistance, effectively resulting in a zero resistance state. This is formally encoded in the theory by an energy

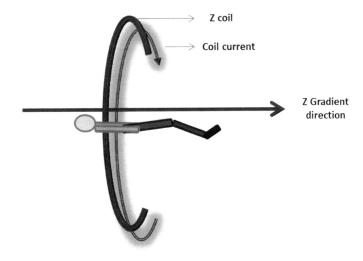

Figure 2.19: Z Gradient coil. Coil is perpendicular to gradient direction in accordance with Faraday-Maxwell Equation.

gap at low (subcritical) temperatures. In superconducting electromagnets, the coil cooling is done through liquid helium and liquid nitrogen and presents a connection to the chemistry of phase transitions, condensed matter and solid state physics, as well as the engineering aspects of attaining and maintaining such low temperatures.

2.14 MRI Summary

Magnetic resonance imaging is an application of quantum mechanics which has revolutionized the practice of medicine. Electrons, protons, and neutrons are magnets by virtue of their spin and orbital angular momentum. Tissue is made of ensembles of such subatomic particles, and can be imaged by sensing their responses to applied magnetic fields. As discussed in this book, a wide array of MRI modalities exist, each of which derives contrast by specific perturbations of the magnetization vector. The potential of MRI in the diagnosis and even the treatment of various diseases is far from fully realized. A quantum level understanding of magnetic resonance technology is essential for further innovation in this field. In

2.14. MRI SUMMARY

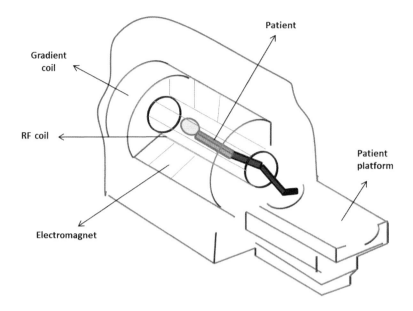

Figure 2.20: MRI machine schematic showing patient, RF coils, gradient coils, and electromagnet.

the next Chapter, we look at how the MRI is being used in the assessment of acute ischemic stroke.

Chapter 2. Magnetic Resonance Imaging

Table 2.1: Physical Constants of Magnetic Resonance Imaging

Quantity	Symbol	Value	Units	u_r
Vacuum light speed	c, c_0	299792458	m s^{-1}	exact
Planck constant	h	$6.62606957(29) \times 10^{-34}$	J s	4.4×10^{-8}
Reduced Planck	$h/2\pi, \hbar$	$1.054571726(47) \times 10^{-34}$	J s	4.4×10^{-8}
Rydberg constant	R_∞	$10973731.568539(55)$	m^{-1}	5.0×10^{-12}
Rydberg energy	R_y, hcR_∞	$13.60569253(30)$	eV	2.0×10^{-8}
Bohr radius	a_0	$0.52917721092(17) \times 10^{-10}$	m	3.2×10^{-10}
Bohr magneton	μ_B	$927.400968(20) \times 10^{-26}$	J T^{-1}	2.2×10^{-8}
Nuclear magneton	μ_N	$5.05078353(11) \times 10^{-27}$	J T^{-1}	2.2×10^{-8}
Electron g-factor	g_e	$-2.00231930436153(53)$		2.6×10^{-13}
Proton g-factor	g_p	$5.585694713(46)$		8.2×10^{-9}
Electron gyromagnetic ratio	γ_e	$1.760859708(39) \times 10^{11}$	s^{-1} T^{-1}	2.2×10^{-8}
	$\gamma_e/2\pi$	$28024.95266(62)$	MHz T^{-1}	2.2×10^{-8}
Proton gyromagnetic ratio	γ_p	$2.675222005(63) \times 10^8$	s^{-1} T^{-1}	2.4×10^{-8}
	$\gamma_p/2\pi$	$42.5774806(10)$	MHz T^{-1}	2.4×10^{-8}
Electron magnetic moment	μ_e	$-928.476430(21) \times 10^{-26}$	J T^{-1}	2.2×10^{-8}
Proton magnetic moment	μ_p	$1.41060743(33) \times 10^{-26}$	J T^{-1}	2.4×10^{-8}
Elementary charge	e	$1.602176565(35) \times 10^{-19}$	C	2.2×10^{-8}
Electric constant	ϵ_0	$8.854187817... \times 10^{-12}$	F m^{-1}	exact
Electron mass	m_e	$9.10938291(40) \times 10^{-31}$	Kg	4.4×10^{-8}
Proton mass	m_p	$1.672621777(74) \times 10^{-27}$	Kg	4.4×10^{-8}
Neutron mass	m_n	$1.674927351(74) \times 10^{-27}$	Kg	4.4×10^{-8}
m_p to m_e ratio	m_p/m_e	$1836.15267245(75)$		4.1×10^{-10}
Avogadro constant	N_A	$6.02214129(27) \times 10^{23}$	mol^{-1}	4.4×10^{-8}
Molar gas const	R	$8.3144621(75)$	J mol^{-1} K^{-1}	9.1×10^{-7}
Boltzmann constant	$R/N_A, k$	$1.3806488(13) \times 10^{-23}$	J K^{-1}	9.1×10^{-7}
Electron volt	$(J/C)e$, eV	$1.602176565(35) \times 10^{-19}$	J	2.2×10^{-8}

2.14. MRI SUMMARY

Table 2.2: Gyromagnetic Ratios of some Biologically-Relevant Nuclei

Element	Nuclei	γ_n (10^6 rad s^{-1} T^{-1})	$\gamma_n/2\pi$ (MHz T^{-1})
Hydrogen	^{1}H	267.513	42.576
Deuterium	D, ^{2}H	41.065	6.536
Helium	^{3}He	-203.789	-32.434
Lithium	^{7}Li	103.962	16.546
Carbon	^{13}C	67.262	10.705
Nitrogen	^{14}N	19.331	3.007
Nitrogen	^{15}N	-27.116	-4.316
Oxygen	^{17}O	-36.264	-5.772
Fluorine	^{19}F	251.662	40.053
Sodium	^{23}Na	70.761	11.262
Phosphorus	^{31}P	108.291	17.235

Chapter 3

Acute Ischemic Stroke

If the term education may be understood in so large a sense as to include all that belongs to the improvement of the mind, either by the acquisition of the knowledge of others or by increase of it through its own exertions, we learn by them what is the kind of education science offers to man. It teaches us to be neglectful of nothing — not to despise the small beginnings, for they precede of necessity all great things in the knowledge of science, either pure or applied. *

—Michael Faraday

---

*'Science as a Branch of Education', Lecture to the Royal Institution, June 11th 1858. Reprinted in *The Mechanics Magazine* (1858), **49**, 11.

CHAPTER 3. ACUTE ISCHEMIC STROKE

In this chapter, we review the MRI modalities used in the assessment of acute ischemic stroke.

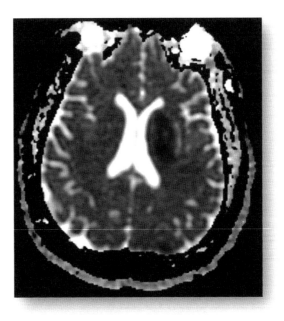

Figure 3.1: ADC map showing hypointensity in the left periventricular MCA distribution, consistent with acute ischemic stroke. Note the marked hyperintensity of the vitreous and cerebrospinal fluid. Both are fluid-filled cavities and have a much higher diffusion coefficient than tissue.

3.1 Diffusion-Weighted Imaging

The diffusion-weighted MRI (DWI) modality exploits the Brownian motion of water molecules within a sample, and the relative phase shift in moving water versus stationary water. The Apparent Diffusion Coefficient (ADC) is a parameter which can be mapped to provide diagnostic information. In acute ischemic stroke, cytotoxic cellular injury results in axonal edema and a subsequent decrease in Brownian motion. This is manifested as a hyperintensity on DWI and a hypointensity on the corresponding ADC map. Figure (3.1)

3.1. DIFFUSION-WEIGHTED IMAGING

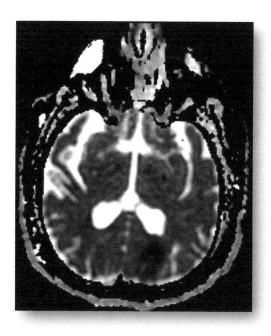

Figure 3.2: ADC map showing hypointensity in the PCA distribution, consistent with a parieto-occipital acute ischemic stroke

is an ADC map showing hypointensity in the left MCA distribution, consistent with acute ischemic stroke. And Figure (3.2) shows an ADC map with a left parieto-occipital hypointensity reflecting left posterior circulation acute ischemic stroke. Figure (3.3) is a DWI showing hyperintensity in the left MCA and PCA distributions, also consistent with acute ischemic stroke in that distribution. In the subacute setting, the ADC may normalize and even increase likely due to ischemia-associated remodeling and loss of structural integrity. The differential for hyperintensity on DWI includes hemorrhagic stroke, traumatic brain injury, multiple sclerosis, and brain abscesses [66, 12, 99, 25]. DWI has long been shown in animal models to be more efficacious than T_2-weighted imaging for the early detection of transient cerebral ischemia [78, 83, 29]. Additionally, DWI has been shown to be highly efficacious in the early detection of acute subcortical infarctions [104]. DWI in con-

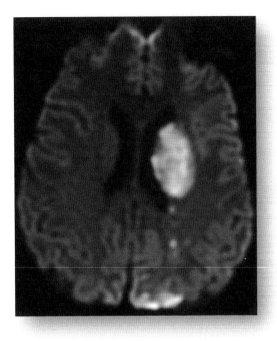

Figure 3.3: DWI showing hyperintensity in the left PCA and MCA distributions, consistent with acute ischemic stroke.

junction with echo-planar imaging has been shown to effectively discriminate between high grade (high cellularity) and low grade (low cellularity) gliomas [108]. Of note, in spite of the relatively high specificity and sensitivity of DWI in the early detection of acute ischemic stroke, there is a small subset of stroke patients who evade DWI detection in spite of clinically evident stroke-like neurological deficits [3]. DWI tractography or diffusion tensor imaging is a form of principal component analysis in which the dominant eigendirection is used to determine the path of an axonal tract in a given voxel. This method has shown potential for further elucidation of neuronal pathways in the brain.

3.2 Perfusion-Weighted Imaging

Perfusion-weighted imaging (PWI) is an ordinary differential equations compartment model of blood perfusion through organs. An arterial input function (AIF) is determined as input into the particular chosen model. The AIF is an impulse, and the model can be described by a summation of the impulse response or Green's function. The computed output are perfusion parameters such as mean transit time, time to peak, cerebral blood flow rate, and cerebral blood volume. There are various protocols for the perfusion conditions. For instance the dynamic susceptibility contrast imaging method uses gadolinium contrast to gather local changes in T_2* signal in surrounding tissue. T_2* is a type of T_2 in which static dephasing effects are not explicitly RF-canceled, therefore dephasing from magnetic field inhomogeneities and susceptibility effects contribute to a tighter FID envelope (more rapid decay; $T_2* < T_2$) than is seen in T_2. Other PWI protocols include arterial spin labeling and blood oxygen level dependent labeling. Not surprisingly, PWI results vary with the choice of perfusion model and computational method with which it is implemented [121, 109, 42, 93].

3.3 Combined DWI and PWI

Both PWI and DWI are efficacious in the early detection of cerebral ischemia and correlate well with various stroke quantitation scales [112, 7, 97]. Furthermore, the combination of diffusion and perfusion-weighted imaging allows for assessment of the ischemic penumbra, that watershed region at the boundary of infarcted and non-infarcted tissue [85, 103, 120, 102]. It represents tissue which has suffered some degree of ischemia, but remains viable and possibly salvageable by expedient intervention with thrombolytic therapy [73, 92, 19], hypothermia, blood pressure elevation, or other experimental methods under study. In the infarcted region, thrombosis and autoregulation of circulation results in a drastic decrease in perfusion. The infarction is itself a result of decreased perfusion, and the overall process can be described by a positive feedback control system. Diffusion is simultaneously decreased due to tissue ischemia, hence no PWI-DWI mismatch occurs. In the penumbra however, the tissue is hypoperfused, yet since still viable, and has yet to sustain sufficient cytotoxic axonal damage to manifest a significant decrease in diffusion. The registration of both imaging modalities therefore demonstrates a diffusion-perfusion mismatch

CHAPTER 3. ACUTE ISCHEMIC STROKE

at the penumbra. In set-theoretic terms, the brain territory Δ with compromised diffusion is a proper subset of the territory Ψ with compromised perfusion, and the relative complement of Δ in Ψ is called the *ischemic penumbra* Λ,

$$\Lambda = \Delta^c \cap \Psi = \Psi \setminus \Delta. \tag{3.1}$$

One caveat to PWI-DWI mismatch assessment is that when done subjectively by the human eye, it can be demonstrably unreliable [27]. Hence development of quantitative metrics and learning algorithms are needed in this area. For instance, generalized linear models have been used and shown promise [120].

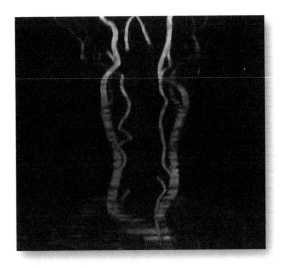

Figure 3.4: 2D time-of-flight fast spoiled gradient echo sequence (FSPGR) image of the carotids.

3.4 Magnetic Resonance Spectroscopy

Magnetic resonance spectroscopy (MRS) is being used in attempts to reliably measure the magnetic resonance of metabolites whose concentrations change in the setting of acute ischemic stroke. Lactate (LAC) levels are well known to increase in ischemia, and is therefore a candidate in development, while N-acetyl aspartate

3.4. MAGNETIC RESONANCE SPECTROSCOPY

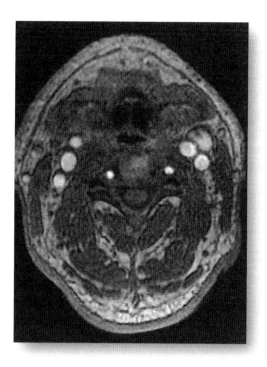

Figure 3.5: 3D time-of-flight image of the carotid vasculature. Axial section of neck shown.

(NAA) levels have been shown to decrease in acute stroke, and are used in proton spectroscopy studies. Most studies confirming the ischemia-associated elevation in LAC and decrease in NAA have been done in the setting of hypoxic-ischemic encephalopathy in newborns [21, 11, 71]. In addition to lactate, α-Glx and glycine have also been shown to be increased in the asphyxiated neonate [68]. Phosphorus magnetic resonance spectroscopy detects changes in the resonance spectrum of energy metabolites, and has also shown prognostic significance in hypoxic-ischemic brain injury [4, 94]. In their current form, MRS methods are less sensitive and practical for the assessment of acute ischemic stroke than the other magnetic resonance modalities discussed here. However, MRS is rapidly finding a place in the routine clinical assessment of hypoxic-ischemic encephalopathy in the asphyxiated neonate.

40 CHAPTER 3. ACUTE ISCHEMIC STROKE

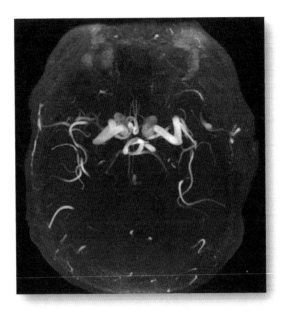

Figure 3.6: MRA showing an axial view of the internal carotids and Circle-of-Willis

Some neonatal intensive care units, for instance, now conduct MRS studies on all neonates below a certain threshold weight.

3.5 BOLD MRI

The blood oxygen level-dependent (BOLD) MRI is a magnetic resonance imaging method that derives contrast from the difference in magnetic properties of oxygenated versus deoxygenated blood. Hemoglobin is the molecule which carries oxygen in the blood, and is located in red blood cells. The oxygen-bound form of hemoglobin is called oxyhemoglobin, while the oxygen-free form is called deoxyhemoglobin. Deoxyhemoglobin is paramagnetic and in the presence of an applied magnetic field assumes a relatively higher magnetic dipole moment than oxyhemoglobin which is a diamagnetic molecule. Deoxygenated blood has a higher concentration of deoxyhemoglobin than oxyhemoglobin, and this difference is reflected in the MRI signal. T_2*-based BOLD MRI has shown promise in lo-

3.5. BOLD MRI

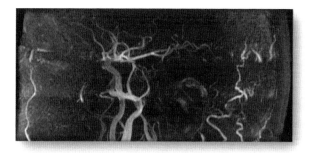

Figure 3.7: MRA lateral view showing internal carotids ascend into the Circle-of-Willis and give rise to the middle cerebral arteries. Outline of the ear is visible posterior to the internal carotids. Branches of the external carotid arteries are also seen as the facial artery anteriorly, and the arteries of the scalp posteriorly.

calizing the penumbra in acute ischemic stroke [46, 41]. It is also used in functional MRI (fMRI), which is based on the principle that active brain areas have higher resource (e.g. oxygen, glucose) demands and higher waste output [47]. In this context, BOLD has been used to study the brain's behavior during sensorimotor recovery following acute ischemic stroke. Specifically, the coupling between BOLD and electrical neuroactivity has shed some light on the still poorly understood process of spontaneous motor recovery following a stroke [16, 20, 56, 98, 28]. BOLD MRI results can be affected by baseline circulatory status, and therefore in the research setting, near-infrared spectroscopy can be used as a control or as an alternative modality in the clinic [106, 84]. BOLD MRI has been used in assessing the brain's response to hypercapnia [114, 6]. Hypercapnia can cause changes in multiple variables in the brain such as cerebral blood flow, oxygen consumption rate, cerebral blood volume, arterial oxygen concentration, and red blood cell volume fraction. However, hypercapnia-associated cerebrovascular reactivity has been strongly correlated with arterial spin labeling and other perfusion surrogates, suggesting that the brain's reaction to hypercapnia is dominated by changes in cerebral blood flow [70].

42 CHAPTER 3. ACUTE ISCHEMIC STROKE

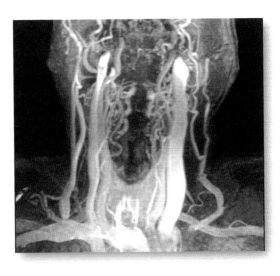

Figure 3.8: MRA of the neck showing the carotid vasculature.

3.6 Magnetic Resonance Angiography

MRA is a set of magnetic resonance-based techniques for imaging the circulatory system. Figures (3.6) and (3.7) show MRA images of the Circle-of-Willis. MRA techniques can be broadly categorized into flow-dependent and flow-independent groups. The flow-dependent methods derive contrast from the motion of blood in vasculature relative to the static state of surrounding tissue [35]. Two currently well-known and used examples of flow dependent methods are: (i) Phase contrast MRA, (PC MRA) [36, 40, 95] and (ii) Time-of-Flight MRA (TOF MRA). The TOF MRA images can be acquired in either two dimensional (2D TOF) or three dimensional (3D TOF) formats. Figure (3.4) shows a 2D TOF fast spoiled gradient echo sequence (FSPGR) image of the carotids, and Figure (3.5) shows a an axial section of a 3D TOF image of the carotids. PC MRA exploits differences in spin phase of moving blood relative to static surrounding tissue, while TOF MRA exploits the difference in excitation pulse (B_1) exposure of flowing blood relative to static surrounding tissue. This difference occurs because flowing blood spends less time in the field of exposure, and as a result is less spin-saturated then the surrounding tissue. This

3.6. MAGNETIC RESONANCE ANGIOGRAPHY 43

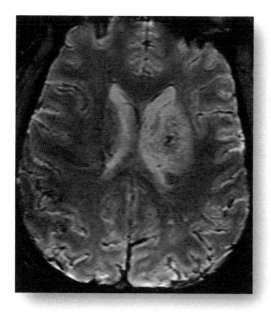

Figure 3.9: SWI showing signal in the left MCA distribution consistent with acute ischemic infarct. Note the left periventricular region of hyperintensity. Its hypointense center is consistent with the ischemic core.

decreased spin-saturation translates into higher intensity signals on spin-echo sequences.

Flow-independent methods exploit inherent differences in magnetic properties of blood relative to surrounding tissue. For instance, fresh blood imaging is a method that exploits the longer T_2 time constant in blood relative to surround. This method finds specific utility in cardiac and cardio-cerebral imaging by using fast spin echo sequences which can exploit the spin saturation differences between systole and diastole [79]. Other examples include susceptibility weighted imaging (SWI) and four dimensional dynamic MRA (4D MRA). SWI derives contrast from magnetic susceptibility differences between blood and surround [50], while 4D MRA uses time-dependent bit-mask subtraction after injec-

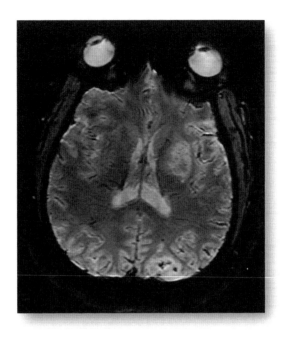

Figure 3.10: SWI showing signal in the left PCA and MCA distributions consistent with acute ischemic infarct. Note the bull's eye pattern of hyperintensity surrounding some central foci of hypointensity in the occipital lesion.

tion of Gadolinium-DPTA or some other pharmacological contrast agent [122]. Figures (3.9) and (3.10) show SWI consistent with acute ischemic infarction in the left PCA and MCA distributions. Of note, in the sense that 4D MRA exploits the time interval between injection and initial image acquisition, it is arguably more flow-dependent than the other methods mentioned here in that category.

Chapter 4

The Hydrogen Atom

Although we know nothing of what an atom is, yet we cannot resist forming some idea of a small particle, which represents it to the mind; and though we are in equal if not greater, ignorance of electricity, so as to be unable to say whether it is a particular matter or matters, or mere motion of ordinary matter, or some third kind of power or agent, yet there is an immensity of facts which justify us in believing that the atoms of matter are in some way endowed or associated with electrical powers, to which they owe their most striking qualities, and amongst them their mutual chemical affinity.[*]

—Michael Faraday

---

[*]'On the absolute quantity of Electricity associated with the particles or atoms of Matter'. In: Dr. Faraday's Experimental Researches in Electricity. In: *Philosophical Transactions of the Royal Society of London* Part I (1834) chapter 13 section 852.

Hydrogen is the most commonly imaged nucleus in magnetic resonance imaging, due largely to the great abundance of water in biological tissue, and to the high gyromagnetic ratio of hydrogen. Furthermore, hydrogen is the only atom whose schrödinger eigenvalue problem has been exactly solved. Other hydrogenic atoms involve screened potentials and require iterative approximation eigenvalue problem solvers such as the Hartree-Fock scheme and its variants. In what follows we review the electronic orbital configuration of hydrogen.

4.1 Electronic Orbital Configuration

In this section we focus on the electron. Specifically that lone electron in the orbit of the hydrogen atom. We make this choice because its simplicity allows us review the attributes of quantum orbital angular momentum and spin, in a context which is not only real, but also highly relevant to magnetic resonance imaging, which as noted targets the hydrogen atom. The electron in ^{1}H has both position and angular momentum. The Hamiltonian commutes with the squared orbital angular momentum operator, therefore angular momentum eigenstates are also energy eigenstates. Of note, the position and angular momentum observables do not commute. This is equivalent to Heisenberg's uncertainty principle. Hence definite energy or orbital angular momentum states, are represented in the position basis as probability density functions, which are synonymous with electron clouds or atomic orbitals.

Orbital angular momentum is the momentum possessed by a body by virtue of circular motion, as along an orbit. In quantum mechanics, the square of the orbital angular momentum is a quantized quantity, which can take on only certain discrete allowed values. This reality and discreteness of the orbital momentum eigenvalues is a manifestation of the spectral theorem of normal bounded operators; given the hermiticity and boundedness of the quantum orbital angular momentum operator over a specified finite volume such as the radius of an atom; which is itself a manifestation of the negative energy of the electron in orbit yielding a so called *bound state*. The orbital momentum states can be indirectly represented by the spherical harmonic functions to be derived below. The spherical harmonics are the eigenfunctions of the squared quantum orbital angular momentum operator in spherical coordinates. Their product with the radial wave functions yield the wave function, ψ, of the hydrogen's electron. Where in accordance with

4.1. ELECTRONIC ORBITAL CONFIGURATION

the principal postulate of quantum mechanics, $\psi^2(r,\theta,\phi)$ is the probability of finding the electron at a given point (r,θ,ϕ).

In the time-independent case, the Hamiltonian defines the Schrödinger equation as follows,

$$H\psi = E\psi. \tag{4.1}$$

The orbital angular momentum states are the angular portion of the solution, ψ, of the schrödinger equation in spherical coordinates. Substituting the expression for the Hamiltonian of a single non-relativistic particle into Equation (4.1) yields,

$$-\frac{\hbar^2}{2m}\nabla^2\psi(\mathbf{r}) + V(\mathbf{r})\psi(\mathbf{r}) = E\psi(\mathbf{r}), \tag{4.2}$$

where $\mathbf{r} \in \Re^3$. In spherical coordinates the above equation becomes,

$$-\frac{\hbar^2 r^{-2}}{2m\sin\theta}\left[\sin\theta\frac{\partial}{\partial r}\left(r^2\frac{\partial}{\partial r}\right) + \frac{\partial}{\partial \theta}\left(\sin\theta\frac{\partial}{\partial \theta}\right) + \frac{1}{\sin\theta}\frac{\partial^2}{\partial \phi^2}\right]\psi(r) + V(r)\psi(r)$$
$$= E\psi(r), \tag{4.3}$$

where m denotes mass. Invoking the method of separation of variables, we assume existence of a solution of the form $\psi(r) = R(r)Y(\theta,\phi)$. We illustrate the method on the Laplace equation which can be interpreted as a homogeneous form of the schrödinger equation, i.e. one for which $V(r) = E = 0$,

$$\frac{Y}{r^2}\frac{\partial}{\partial r}\left(r^2\frac{\partial}{\partial r}\right)R(r) + \frac{R}{r^2\sin\theta}\frac{\partial}{\partial \theta}\left(\sin\theta\frac{\partial}{\partial \theta}\right)Y(\theta,\phi) +$$
$$\frac{R}{r^2\sin^2\theta}\frac{\partial^2 Y}{\partial \phi^2}$$
$$= 0. \tag{4.4}$$

Multiplying through by r^2/RY yields,

$$\frac{1}{R}\frac{\partial}{\partial r}\left(r^2\frac{\partial}{\partial r}\right)R(r) + \frac{1}{Y\sin\theta}\frac{\partial}{\partial\theta}\left(\sin\theta\frac{\partial}{\partial\theta}\right)Y(\theta,\phi) +$$
$$\frac{1}{Y\sin^2\theta}\frac{\partial^2 Y}{\partial\phi^2}$$
$$= 0, \quad (4.5)$$

which can be separated as,

$$\frac{1}{R}\frac{\partial}{\partial r}\left(r^2\frac{\partial}{\partial r}\right)R(r) = \xi \quad (4.6)$$

and

$$\frac{1}{Y\sin\theta}\frac{\partial}{\partial\theta}\left(\sin\theta\frac{\partial}{\partial\theta}\right)Y(\theta,\phi) + \frac{1}{Y\sin^2\theta}\frac{\partial^2 Y}{\partial\phi^2} = -\xi. \quad (4.7)$$

We can again assume separability in the form $Y(\theta,\phi) = \Theta(\theta)\Phi(\phi)$, and substitute into Equation (4.7) above to get,

$$\frac{1}{\Theta\sin\theta}\frac{\partial}{\partial\theta}\left(\sin\theta\frac{\partial}{\partial\theta}\right)\Theta(\theta) + \frac{1}{\Phi\sin^2\theta}\frac{\partial^2 \Phi}{\partial\phi^2} = -\xi. \quad (4.8)$$

From which we extract the following two separated equations,

$$\frac{1}{\Phi}\frac{\partial^2 \Phi}{\partial\phi^2} = -m^2 \quad (4.9)$$

and

$$\xi\sin^2\theta + \frac{\sin\theta}{\Theta}\frac{\partial}{\partial\theta}\left(\sin\theta\frac{\partial\Theta}{\partial\theta}\right) = m^2. \quad (4.10)$$

Our homogeneous equation has therefore been separated into three ordinary differential equations: Equations (4.6) for the radial part, Equation (4.9) for the azimuthal part, and Equation (4.10) for the polar part. Solutions to the radial equation are of the form,

$$R(l) = Ar^l + Br^{-(l+1)} \quad (4.11)$$

where A and B are constant coefficients, l is a non-negative integer such that $l \geq |m|$, where m is an integer on the right hand side of the azimuthal and polar equations. A regularity constraint

4.1. ELECTRONIC ORBITAL CONFIGURATION

at the poles of the sphere yield a Sturm-Liouville problem which in turn mandates the form $\xi = l(l+1)$. The coefficient A is often set to zero to admit only solutions which vanish at infinity. However, this choice is application-specific, and for certain applications it may be appropriate to instead set $B = 0$.

Solutions to the azimuthal equation are of the form,

$$Ce^{-im\phi} + De^{+im\phi}, \tag{4.12}$$

where C and D are constant coefficients and e is the base of the natural logarithm. To obtain solutions to the polar equation, we proceed in a number of steps. First we substitute $\cos\theta \mapsto x$, and recast Equation (4.10) into,

$$\sin^2\theta \frac{\partial^2 \Theta}{\partial \theta^2} + \sin\theta \cos\theta \frac{\partial \Theta}{\partial \theta} + l(l+1)\Theta \sin^2\theta - m^2 \Theta = 0. \tag{4.13}$$

Next we compute the derivatives of Θ under the transformation $x = \cos\theta$. Employing the chain rule yields,

$$\frac{d\Theta}{d\theta} = \frac{d\Theta}{dx}\frac{dx}{d\theta} = -\sin\theta \frac{d\Theta}{dx} \tag{4.14}$$

and

$$\frac{d^2\Theta}{d\theta^2} = \frac{d}{d\theta}\left(-\sin\theta \frac{d\Theta}{dx}\right) = -\cos\theta \frac{d\Theta}{dx} - \sin\theta \frac{d}{d\theta}\frac{d\Theta}{dx}$$
$$= \sin^2\theta \frac{d^2\Theta}{dx^2} - \cos\theta \frac{d\Theta}{dx}. \tag{4.15}$$

A substitution of the derivatives into the polar equation yields,

$$\sin^2\theta \left(\sin^2\theta \frac{d^2\Theta}{dx^2} - \cos\theta \frac{d\Theta}{dx}\right) + \sin\theta \cos\theta \left(-\sin\theta \frac{d\Theta}{dx}\right)$$
$$+ l(l+1)\Theta \sin^2\theta - m^2\Theta = 0. \tag{4.16}$$

Dividing through by $\sin^2\theta$ and changing variables via $\cos\theta \mapsto x$ and $\Theta \mapsto y$, yields the associated ($m \neq 0$) Legendre differential equation,

$$(1-x^2)\frac{\partial^2 y}{\partial x^2} - 2x\frac{\partial y}{\partial x} + \left(l(l+1) - \frac{m^2}{1-x^2}\right)y = 0, \tag{4.17}$$

whose solutions are given by

$$P_l^m(x) = \frac{(-1)^m}{2^l l!}(1-x^2)^{m/2}\frac{d^{l+m}}{dx^{l+m}}(x^2-1)^l. \quad (4.18)$$

The solution $Y(\phi,\theta) = \Phi(\phi)\Theta(\theta)$ to the angular portion of the Laplace equation is therefore of the form,

$$Y_{l,m}(\theta,\phi) = NP_l^m(\cos\theta)e^{im\phi}, \quad (4.19)$$

where N is a normalization factor given by,

$$N = \sqrt{\frac{(2l+1)}{4\pi}\frac{(l-1)!}{(l+1)!}}, \quad (4.20)$$

and enabling,

$$\int_\Omega Y_{l,m}^*(\Omega)Y_{l,m}(\Omega) = 1, \quad (4.21)$$

where $\Omega = (\phi,\theta)$ is solid angle. $Y_{l,m}$ is called the spherical harmonic function of degree l and order m, and is discussed some more in the following section. Figures (4.1) to (4.10) are plots of a sampling of spherical harmonics with $1 \leq l \leq 5$.

For the electronic configuration of the hydrogen atom, the electron experiences a potential V(r) due to the proton. V(r) is the coulomb potential given by,

$$V(r) = -\frac{e^2}{4\pi\epsilon_0 r} \quad (4.22)$$

Substituting this into the generic schrödinger equation yields,

$$-\frac{\hbar^2}{2m}\nabla^2\psi(r) - \frac{e^2}{4\pi\epsilon_0 r}\psi(r) = E\psi(r), \quad (4.23)$$

which in spherical coordinates is,

$$-\frac{\hbar^2 r^{-2}}{2m\sin\theta}\left[\sin\theta\frac{\partial}{\partial r}\left(r^2\frac{\partial}{\partial r}\right) + \frac{\partial}{\partial\theta}\left(\sin\theta\frac{\partial}{\partial\theta}\right) + \frac{1}{\sin\theta}\frac{\partial^2}{\partial\phi^2}\right]\psi(r)$$

$$-\frac{e^2}{4\pi\epsilon_0 r}\psi(r) = E\psi(r), \quad (4.24)$$

4.1. ELECTRONIC ORBITAL CONFIGURATION

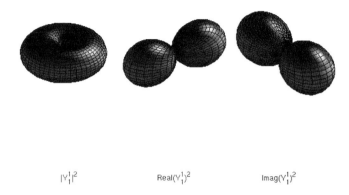

|Y₁¹|²   Real(Y₁¹)²   Imag(Y₁¹)²

Figure 4.1: Spherical Harmonics: Y_1^1

where m is given by,

$$m = \frac{m_p m_e}{m_p + m_e}, \qquad (4.25)$$

and is the two-body reduced mass m_p of the proton and m_e of the electron. Equation (4.24) above is an eigenvalue problem which after some rearranging, we can solve using the same separation of variables method illustrated above. We recast as,

$$\left[\frac{\partial}{\partial r}\left(r^2 \frac{\partial}{\partial r}\right) + \frac{2m}{\hbar^2}\left(\frac{re^2}{4\pi\epsilon_0} + Er^2\right)\right] R(r)Y(\theta, \psi)$$
$$+ \left[\frac{1}{\sin\theta}\frac{\partial}{\partial\theta}\left(\sin\theta \frac{\partial}{\partial\theta}\right) + \frac{1}{\sin^2\theta}\frac{\partial^2}{\partial\phi^2}\right] R(r)Y(\theta, \psi)$$
$$= 0. \qquad (4.26)$$

Dividing through by $R(r)Y(\theta, \phi)$ and splitting the operator, we get,

$$\frac{1}{Y(\Omega)}\left[\frac{1}{\sin\theta}\frac{\partial}{\partial\theta}\left(\sin\theta \frac{\partial}{\partial\theta}\right) + \frac{1}{\sin^2\theta}\frac{\partial^2}{\partial\phi^2}\right] Y(\Omega) = -l(l+1) \quad (4.27)$$

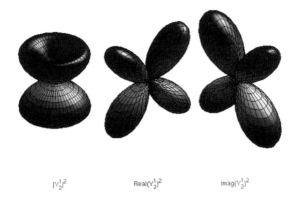

Figure 4.2: Spherical Harmonics: Y_2^1

and

$$\frac{1}{R(r)}\frac{\partial}{\partial r}\left(r^2\frac{\partial}{\partial r}\right)R(r) + \frac{2m}{\hbar^2}\left[\frac{re^2}{4\pi\epsilon_0} + Er^2\right] = l(l+1). \quad (4.28)$$

Equation (4.27) is exactly the same as Equation (4.8) which we solved above, and whose solutions are the spherical harmonic functions, $Y_{l,m}$. Equation (4.28) above is isomorphic to a generalized Laguerre differential equation, whose solutions are related to the associated Laguerre polynomials. As mentioned in the case of Laplace's equation above, regularity conditions at the boundary prescribe the separation constant $l(l+1)$. To transform Equation (4.28) into a generalized Laguerre equation, we recast into the following form,

$$\frac{\partial}{\partial r}\left(r^2\frac{\partial}{\partial r}\right)R(r) + \left[\frac{2mre^2}{4\pi\epsilon_0\hbar^2} + \frac{2mEr^2}{\hbar^2} - l(l+1)\right]R(r) = 0. \quad (4.29)$$

We then proceed with the following sequence of substitutions:

4.1. ELECTRONIC ORBITAL CONFIGURATION

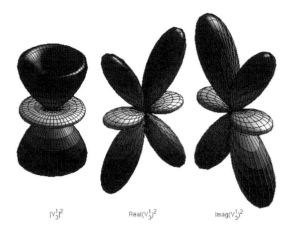

Figure 4.3: Spherical Harmonics: Y_3^1

$$R(r) \mapsto \frac{y(r)}{r},$$
$$-\frac{2mE}{\hbar^2} \mapsto \frac{\beta^2}{4},$$
$$r \mapsto \frac{x}{\beta}. \quad (4.30)$$

Executing the first substitution above transforms Equation (4.29) into,

$$\frac{d^2 y(r)}{dr^2} + \left[\frac{2me^2}{4\pi r \epsilon_0 \hbar^2} + \frac{2mE}{\hbar^2} - \frac{l(l+1)}{r^2} \right] y(r) = 0. \quad (4.31)$$

Next, the second substitution transforms the above into,

$$\frac{d^2 y(r)}{dr^2} + \left[\frac{2me^2}{4\pi r \epsilon_0 \hbar^2} - \frac{\beta^2}{4} - \frac{l(l+1)}{r^2} \right] y(r) = 0. \quad (4.32)$$

And finally, the third substitution transforms the above into,

$$\beta^2 \frac{d^2 y(x)}{dx^2} + \left[\frac{2\beta me^2}{4\pi x \epsilon_0 \hbar^2} - \frac{\beta^2}{4} - \frac{\beta^2 l(l+1)}{x^2} \right] y(x) = 0. \quad (4.33)$$

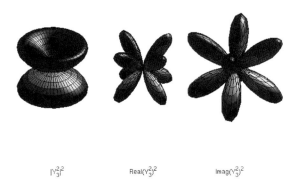

Figure 4.4: Spherical Harmonics: Y_3^2

Dividing through by β^2 yields,

$$\frac{d^2y(x)}{dx^2} + \left[-\frac{1}{4} + \frac{2me^2}{4\pi\beta\epsilon_0\hbar^2 x} - \frac{l(l+1)}{x^2}\right]y(x) = 0. \quad (4.34)$$

Next, we make the substitutions,

$$l(l+1) \mapsto \frac{k^2 - 1}{4} \quad (4.35)$$

and

$$\frac{2me^2}{4\pi\beta\epsilon_0\hbar^2} \mapsto \frac{2j + k + 1}{2}, \quad (4.36)$$

which transform Equation (4.34) into the following generalized Laguerre equation

$$(y_j^k)''(x) + \left[-\frac{1}{4} + \frac{2j + k + 1}{2x} - \frac{k^2 - 1}{4x^2}\right]y_j^k(x) = 0, \quad (4.37)$$

whose solutions are readily verified to be the following associated Laguerre functions,

$$y_j^k(x) = e^{-x/2}x^{(k+1)/2}L_j^k(x), \quad (4.38)$$

4.1. ELECTRONIC ORBITAL CONFIGURATION

|Y$_3^3$|^2 Real(Y$_3^3$)^2 Imag(Y$_3^3$)^2

Figure 4.5: Spherical Harmonics: Y_3^3

where L_j^k are the associated Laguerre polynomials. From Equation (4.35) we see that,

$$k = 2l + 1, \tag{4.39}$$

which we substitute into Equation (4.36) to yield,

$$\frac{2me^2}{4\pi\beta\epsilon_0\hbar^2} = \frac{2j + (2l+1) + 1}{2} = j + l + 1 \equiv n. \tag{4.40}$$

n is the principal (or radial) quantum number. The corresponding energy eigenvalues, E_n, are obtained by substituting β^2 into n^2 and solving for E, to get,

$$E_n = -\frac{\hbar^2}{2ma_0^2 n^2}, \tag{4.41}$$

where a_0 is the Bohr radius and is given by,

$$a_0 = \frac{4\pi\epsilon_0\hbar^2}{m_e e^2}. \tag{4.42}$$

56   CHAPTER 4. THE HYDROGEN ATOM

Figure 4.6: Spherical Harmonics: Y_4^1

The value of E_n in the ground state ($n = 1$) is called the Rydberg constant. Its value is shown in Table (2.1) along with that of other physical constants pertinent to magnetic resonance imaging. The displayed values are from the 2010 Committee for Data on Science and Technology (CODATA) recommendations [81], and u_r is the relative standard uncertainty.

To transform the radial equation solution, Equation (4.38), into a form in terms of radial distance r, and principal and azimuthal quantum numbers n and l, we note the following relations:

From Equation (4.39) we get,

$$\frac{k+1}{2} = \frac{(2l+1)+1}{2} = l+1; \tag{4.43}$$

from Equation (4.40) we get,

$$j + l + 1 \Rightarrow j = n - l - 1; \tag{4.44}$$

and from Equations (4.30) and (4.41) we get,

$$\frac{\beta^2}{4} = -\frac{2mE_n}{\hbar^2} = -\frac{2m}{\hbar^2}\left(-\frac{\hbar^2}{2ma_0^2 n^2}\right)$$
$$= \frac{1}{(a_0 n)^2} \Rightarrow \beta = \frac{2}{a_0 n}. \tag{4.45}$$

4.1. ELECTRONIC ORBITAL CONFIGURATION

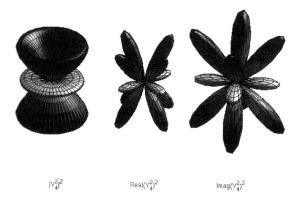

Figure 4.7: Spherical Harmonics: Y_4^2

Therefore,

$$x = \beta r = \frac{2r}{a_0 n}. \tag{4.46}$$

Substituting the above derived expressions of x, j, and k into the associated Laguerre function, Equation (4.38), we get,

$$y_{n-l-1}^{2l+1}(r) = e^{-r/a_0 n}\left(\frac{2r}{a_0 n}\right)^{l+1} L_{n-l-1}^{2l+1}\left(\frac{2r}{a_0 n}\right). \tag{4.47}$$

Next we make the substitution $R = y(r)/r$, which yields the radial solutions of the hydrogen atom,

$$R(r) = Ce^{-r/a_0 n}\left(\frac{2r}{a_0 n}\right)^{l} L_{n-l-1}^{2l+1}\left(\frac{2r}{a_0 n}\right), \tag{4.48}$$

where the constant $C = \frac{2}{a_0 n}$. Following multiplication by the spherical harmonics and normalization, we obtain the exact solution of the time-independent Schrödinger equation of the hydrogen atom,

$|Y_4^3|^2$ Real$(Y_4^3)^2$ Imag$(Y_4^3)^2$

Figure 4.8: Spherical Harmonics: Y_4^3

$$\langle r,\theta,\phi|n,l,m\rangle = \Psi(r,\theta,\phi) =$$
$$Ne^{-r/na_0}\left(\frac{2r}{na_0}\right)^l L_{n-l-1}^{2l+1}\left(\frac{2r}{na_0}\right) Y_{l,m}(\theta,\phi), \qquad (4.49)$$

where the normalization factor N is given by,

$$N = \sqrt{\left(\frac{2}{na_0}\right)^3 \frac{(n-l-1)!}{2n[(n+1)!]^3}},$$

and the principal, azimuthal, and magnetic quantum numbers (n, l, m) take the following values,

$$n = 1, 2, 3...$$
$$l = 0, 1, 2, ...n-1$$
$$m = -l, -l+1, ...0, ...l-1, l$$

4.2. SPHERICAL HARMONICS AND RADIAL WAVES

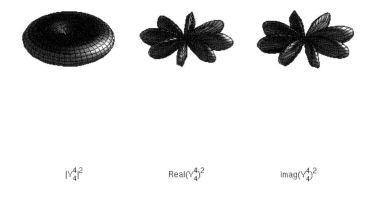

Figure 4.9: Spherical Harmonics: Y_4^4

4.2 Spherical Harmonics and Radial Waves

In this section, we review the spherical harmonics and the radial wave functions. As noted above, the spherical harmonics, $Y_l^{m_l}$, are characterized by an azimuthal and a magnetic quantum number, l and m_l respectively. Table (4.1) shows the spherical harmonics with $l = 0, 1, 2,$ and 3. Formalism of state representation by the spherical harmonics is as follows,

$$|l, m\rangle = Y_{l,m} \tag{4.50}$$

$$\langle \theta, \phi | l, m \rangle = Y_{l,m}(\theta, \phi) \tag{4.51}$$

where, $0 \leq \theta \leq \pi$ is the polar angle and $0 \leq \phi \leq 2\pi$ is the azimuthal angle. The Spherical harmonic representing rotation quantum numbers $|l, m\rangle$ at position state $|\theta, \phi\rangle$, is given by

$$Y_{l,m}(\theta, \phi) = \sqrt{\frac{(2l+1)}{4\pi} \frac{(l-1)!}{(l+1)!}} P_l^m(\cos\theta) e^{im\phi}. \tag{4.52}$$

Figure 4.10: Spherical Harmonics: Y_5^2

Operator Representation

The spherical harmonics are a complete set of orthonormal functions over the unit sphere. In particular, they are eigenfunctions of the square of the orbital angular momentum operator, L^2. And are thereby representations of the allowed discrete states of angular momentum. L^2 can be arrived at by representing Laplace's equation in spherical coordinates and considering only its angular portion. Classically, angular momentum is:

$$\mathbf{L} = \mathbf{r} \times \mathbf{p}, \tag{4.53}$$

where $\mathbf{r}$ is the position vector and $\mathbf{p}$ is the momentum vector.

By analogy, the quantum orbital angular momentum operator is given by,

$$\mathbf{L} = -i\hbar \left(\mathbf{r} \times \nabla \right), \tag{4.54}$$

where ∇ is the gradient operator, and we have used the momentum operator of quantum mechanics,

$$\mathbf{p} = -i\hbar \nabla. \tag{4.55}$$

In spherical coordinates, L^2 is then represented by,

4.2. SPHERICAL HARMONICS AND RADIAL WAVES

Table 4.1: Spherical Harmonics for $l = 0, 1, 2,$ and 3.

l	m_l	$Y_l^{m_l}$
0	0	$Y_0^0(\theta, \phi) = \frac{1}{2}\sqrt{\frac{1}{\pi}}$
1	-1	$Y_1^{-1}(\theta, \phi) = \frac{1}{2}\sqrt{\frac{3}{2\pi}} \sin\theta e^{-i\phi}$
1	0	$Y_1^0(\theta, \phi) = \frac{1}{2}\sqrt{\frac{3}{\pi}} \cos\theta$
1	1	$Y_1^1(\theta, \phi) = -\frac{1}{2}\sqrt{\frac{3}{2\pi}} \sin\theta e^{i\phi}$
2	-2	$Y_2^{-2}(\theta, \phi) = \frac{1}{4}\sqrt{\frac{15}{2\pi}} \sin^2\theta e^{-2i\phi}$
2	-1	$Y_2^{-1}(\theta, \phi) = \frac{1}{2}\sqrt{\frac{15}{2\pi}} \sin\theta \cos\theta e^{-i\phi}$
2	0	$Y_2^0(\theta, \phi) = \frac{1}{4}\sqrt{\frac{5}{\pi}} (3\cos^2\theta - 1)$
2	1	$Y_2^1(\theta, \phi) = -\frac{1}{2}\sqrt{\frac{15}{2\pi}} \sin\theta \cos\theta e^{i\phi}$
2	2	$Y_2^2(\theta, \phi) = \frac{1}{4}\sqrt{\frac{15}{2\pi}} \sin^2\theta e^{2i\phi}$
3	-3	$Y_3^{-3}(\theta, \phi) = \frac{1}{8}\sqrt{\frac{35}{\pi}} \sin^3\theta e^{-3i\phi}$
3	-2	$Y_3^{-2}(\theta, \phi) = \frac{1}{4}\sqrt{\frac{105}{2\pi}} \sin^2\theta \cos\theta e^{-2i\phi}$
3	-1	$Y_3^{-1}(\theta, \phi) = \frac{1}{8}\sqrt{\frac{21}{\pi}} (5\cos^2\theta - 1)\sin\theta e^{-i\phi}$
3	0	$Y_3^0(\theta, \phi) = \frac{1}{4}\sqrt{\frac{7}{\pi}} (5\cos^3\theta - 3\cos\theta)$
3	1	$Y_3^1(\theta, \phi) = -\frac{1}{8}\sqrt{\frac{21}{\pi}} (5\cos^2\theta - 1)\sin\theta e^{i\phi}$
3	2	$Y_3^2(\theta, \phi) = \frac{1}{4}\sqrt{\frac{105}{2\pi}} \sin^2\theta \cos\theta e^{2i\phi}$
3	3	$Y_3^3(\theta, \phi) = -\frac{1}{8}\sqrt{\frac{35}{\pi}} \sin^3\theta e^{3i\phi}$

$$L^2 = -\hbar^2 \left(\frac{1}{\sin\theta} \frac{\partial}{\partial \theta} \left(\sin\theta \frac{\partial}{\partial \theta} \right) + \frac{1}{\sin^2\theta} \frac{\partial^2}{\partial \phi^2} \right), \quad (4.56)$$

and similarly, L_x, L_y, and L_z are represented as follows,

$$L_x = i\hbar \left(\sin\phi \frac{\partial}{\partial \theta} + \cot\theta \cos\phi \frac{\partial}{\partial \phi} \right), \quad (4.57)$$

$$L_y = i\hbar \left(-\cos\phi \frac{\partial}{\partial \theta} + \cot\theta \sin\phi \frac{\partial}{\partial \phi} \right), \quad (4.58)$$

and

$$L_z = -i\hbar \frac{\partial}{\partial \phi}. \quad (4.59)$$

Table 4.2: Radial Wave Functions for $n = 0, 1, 2,$ and 3.

n	l	R_n^l
1	0	$R_1^0(r) = 2\left(\frac{1}{a_0}\right)^{3/2} e^{-r/a_0}$
2	0	$R_2^0(r) = \left(\frac{1}{2a_0}\right)^{3/2} (2 - r/a_0) e^{-r/2a_0}$
2	1	$R_2^1(r) = \left(\frac{1}{2a_0}\right)^{3/2} \frac{r}{a_0\sqrt{3}} e^{-r/2a_0}$
3	0	$R_3^0(r) = 2\left(\frac{1}{3a_0}\right)^{3/2} \left(1 - \frac{2r}{3a_0} + \frac{2}{27}(r/a_0)^2\right) e^{-r/3a_0}$
3	1	$R_3^1(r) = \left(\frac{1}{3a_0}\right)^{3/2} \frac{4\sqrt{2}}{3} \left(1 - \frac{r}{6a_0}\right) e^{-r/3a_0}$
3	2	$R_3^2(r) = \left(\frac{1}{3a_0}\right)^{3/2} \frac{2\sqrt{2}}{27\sqrt{5}} \left(\frac{r}{a_0}\right)^2 e^{-r/3a_0}$

Given the expressions above for $Y_{l,m}(\theta, \phi)$, L^2, and L_z, it is readily shown that,

$$L^2 |l, m\rangle = \hbar^2 l(l+1) |l, m\rangle, \tag{4.60}$$

and

$$L_z |l, m\rangle = \hbar m |l, m\rangle. \tag{4.61}$$

The square of the radial wave functions is the probability the electron is located a distance r from the nucleus. Table (4.2) shows the radial wave functions for $n = 1, 2,$ and 3.

Commutation Relations and Ladder Operators

$$[L_i, L_j] = i\hbar \epsilon_{ijk} L_k, \tag{4.62}$$

where ϵ_{ijk} is the Levi-Civita symbol given by,

$$\epsilon_{ijk} = \begin{cases} +1 & \text{if (i,j,k) is an even permutation of (1,2,3),} \\ -1 & \text{if (i,j,k) is an odd permutation of (1,2,3),} \\ 0 & \text{if any index is repeated,} \end{cases} \tag{4.63}$$

and i, j, k can have values of x, y, or z.

$$[L^2, L_j] = 0 \quad \text{for } j = x, y, z \tag{4.64}$$

The ladder operators act to increase or decrease the m quantum number of a state and are given by,

4.2. SPHERICAL HARMONICS AND RADIAL WAVES   63

$$L_\pm = L_x \pm iL_y. \tag{4.65}$$

It is readily verified that,

$$L_\pm |l,m\rangle = \hbar\sqrt{(l \mp m)(l+1 \pm m)}|l,m \pm 1\rangle, \tag{4.66}$$

$$[L_z, L_\pm] = \pm\hbar L_\pm, \tag{4.67}$$

and

$$[L_+, L_-] = 2\hbar L_z. \tag{4.68}$$

Chapter 5
Intrinsic Spin

I am no poet, but if you think for yourselves, as I proceed, the facts will form a poem in your minds.[*]

—Michael Faraday

---

[*]Lecture notes of 1858, from *The Life and Letters of Faraday* (1870), H. B. Jones (ed.), Vol. 2, p. 403

Spin is an intrinsic quantum property of elementary particles such as the electron and the quark. Composite particles such as the neutron, proton, and even atoms and molecules also possess spin by virtue of their composition from their elementary particle constituents. The term spin is itself a misnomer, as a physically spinning object about an axis is not sufficient to account for the observed magnetic moments. The mathematics of spin was worked out by Wolfgang Pauli, who either astutely or serendipitously neglected to name it. This was wise, as later insight elucidated spin as an intrinsic quantum mechanical property with no classical correlate.

5.1 Spin Algebra

Spin follows essentially the same mathematics as outlined above for orbital angular momentum. This is by virtue of the isomorphism of the respective Lie Algebras of SO(3) and SU(2) groups. And is elaborated further in Chapter (7) below. One notable difference is that half-integer eigenvalues are allowed for spin, while orbital angular momentum admits only integer eigenvalues. Recall from Chapter (2) that,

$$\mathbf{S}^2|s,m\rangle = \hbar^2 s(s+1)|s,m\rangle,$$

and

$$\mathbf{S}_z|s,m\rangle = \hbar m|s,m\rangle,$$

where $|s,m\rangle$ is a spin eigenstate vector for a particle of spin s and magnetic spin quantum number m, $\mathbf{S}^2$ is the square of the spin, and $\mathbf{S}_z$ is the z-component of spin. In general, $m \in \{-s, -(s-1), ..., s-1, s\}$, and in particular for the electron and proton $s = 1/2$ and $m \in \{-1/2, 1/2\}$. The hermitian operators $\mathbf{S}$ are given by,

$$\mathbf{S}_j = \frac{\hbar}{2}\boldsymbol{\sigma}_j, \tag{5.1}$$

where $j \in \{1,2,3\}$, and $\boldsymbol{\sigma}_j$ are the Pauli spin matrices given by,

$$\boldsymbol{\sigma}_1 = \begin{pmatrix} 0 & 1 \\ 1 & 0 \end{pmatrix}, \quad \boldsymbol{\sigma}_2 = \begin{pmatrix} 0 & -i \\ i & 0 \end{pmatrix}, \quad \boldsymbol{\sigma}_3 = \begin{pmatrix} 1 & 0 \\ 0 & -1 \end{pmatrix}. \tag{5.2}$$

5.1. SPIN ALGEBRA

The spin algebra is encapsulated in the commutation relations as follows,

$$[\mathbf{S}_i, \mathbf{S}_j] = i\hbar\epsilon_{ijk}\mathbf{S}_k, \tag{5.3}$$

$$[\mathbf{S}^2, \mathbf{S}_j] = 0 \quad \text{for } j = x, y, z, \tag{5.4}$$

$$\mathbf{S}_\pm = \mathbf{S}_x \pm i\mathbf{S}_y, \tag{5.5}$$

$$\mathbf{S}_\pm|s,m\rangle = \hbar\sqrt{(s \mp m)(s+1 \pm m)}|s, m \pm 1\rangle, \tag{5.6}$$

$$[\mathbf{S}_z, \mathbf{S}_\pm] = \pm\hbar\mathbf{S}_\pm, \tag{5.7}$$

and

$$[\mathbf{S}_+, \mathbf{S}_-] = 2\hbar\mathbf{S}_z. \tag{5.8}$$

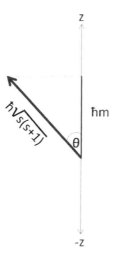

Figure 5.1: Spin up alignment geometry.

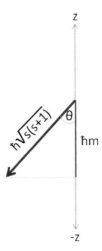

Figure 5.2: Spin down alignment geometry.

5.2 Spin Alignment Geometry

The application of an external magnetic field yields alignment of the spins in either parallel or antiparallel fashion. The extent of parallelism is constrained by the eigenvalues of the $\mathbf{S}^2$ and $\mathbf{S}_z$ operators above, such that from the geometry, the parallel and antiparallel states make angles of $\cos^{-1}(1/\sqrt{3}) \approx 54.7356103172$ degrees with the positive and negative z-axes respectively (assuming $\mathbf{B}_0$ is oriented exclusively along the z-axis). Figures (5.1) and (5.2) illustrate the alignment geometries of the spin up and spin down states respectively. The figures depict the following relationship,

5.2. SPIN ALIGNMENT GEOMETRY

$$\begin{aligned}
\cos\theta &= \frac{\hbar m}{\hbar\sqrt{s(s+1)}} \Rightarrow \\
\theta &= \cos^{-1}\left(\frac{\hbar m}{\hbar\sqrt{s(s+1)}}\right) \\
&= \cos^{-1}\left(\frac{1/2}{\sqrt{1/2(1/2+1)}}\right) \\
&= \cos^{-1}\left(\frac{1}{\sqrt{3}}\right) \\
&\approx 54.7356103172.
\end{aligned} \qquad (5.9)$$

Chapter 6

Clebsch-Gordan Coefficients

The laws of nature, as we understand them, are the foundation of our knowledge in natural things. So much as we know of them has been developed by the successive energies of the highest intellects, exerted through many ages. After a most rigid and scrutinizing examination upon principle and trial, a definite expression has been given to them; they have become, as it were, our belief or trust. From day to day we still examine and test our expressions of them. We have no interest in their retention if erroneous. On the contrary, the greatest discovery a man could make would be to prove that one of these accepted laws was erroneous, and his greatest honour would be the discovery. Neither should there be any desire to retain the former expression— for we know that the new or amended law would be far more productive in results, would greatly increase our intellectual acquisitions, and would prove an abundant source of fresh delight to the mind.[*]

—Michael Faraday

---

[*]Experimental Researches in Chemistry and Physics (1859), p.469. Reprinted from *The Philosophical Transactions of 1821-1857; The Journal of the Royal Institution; The Philosophical Magazine; and Other Publications*

The addition of spin and or of orbital angular momentum refers to the process of determining the net spin and or orbital angular momentum of a system of particles. For example, the spin of a hydrogen atom in its ground state, i.e. the $l = 0$ state with zero orbital angular momentum, constitutes the composite spin of the electron and the proton. Similarly, the nuclear spin, I, is the composition of the spins of protons and neutrons in the nucleus of an atom.

Mathematically, the process is described by a change of basis. A change from the product (uncoupled) space basis $\langle j_1, m_1 | \langle j_2, m_2 |$ to the coupled space basis $\langle j, m |$, where $m = m_1 + m_2$ and $|j_1 - j_2| \leq j \leq j_1 + j_2$. The representation of the coupled space eigenstates in terms of the product space eigenstates are referred to as the *Clebsch-Gordan coefficients*.

$$|j, m\rangle = \sum_{m_1=-j_1}^{j_1} \sum_{m_2=-j_2}^{j_2} C(m_1, m_2, m) |j_1, m_1\rangle |j_2, m_2\rangle, \qquad (6.1)$$

where $C(m_1, m_2, m)$ are the Clebsch-Gordan coefficients and are given by,

$$C(m_1, m_2, m) = \langle j, m | j_1 j_2, m_1 m_2 \rangle, \qquad (6.2)$$

and we have notationally represented the uncoupled product space basis by,

$$|j_1, m_1\rangle |j_2, m_2\rangle := |j_1 j_2, m_1 m_2\rangle. \qquad (6.3)$$

6.1 Addition Algorithm

Here we review a procedural description of the addition of angular momenta. Consider a hydrogen atom ^{1}H in the ground state. The spin contributors are the lone electron in the l=0 shell, and a proton which is the sole constituent of the nucleus. The possible configurations of spin are:

$$|\uparrow\uparrow\rangle, \ |\uparrow\downarrow\rangle, \ |\downarrow\uparrow\rangle, \text{ and } |\uparrow\uparrow\rangle \qquad (6.4)$$

where $\uparrow$ indicates spin $+1/2$ (spin up) and $\downarrow$ indicates spin $-1/2$ (spin down). And where each of the above states are prepared by measuring S_1^2, S_2^2, $S_{z,1}$ and $S_{z,2}$ for each state. We are interested in the combined state of the two particle system, and this requires

6.1. ADDITION ALGORITHM

the quantum mechanical addition of angular momenta, or in this example, spin. The net result is a state $|s, m\rangle$ whose spin can be determined by measuring S^2 and whose z-component of spin can be determined by measuring S_z, where $S = S_1 + S_2$ and $S_z = S_{z,1} + S_{z,2}$. It follows that $m = m_1 + m_2$, and therefore the possible values of m are 1, 0, 0, -1. This suggests a triplet state (s=1) and a singlet state (s=0). The triplet state corresponds to m=1,0,-1, and is given by,

$$\begin{aligned} |1,1\rangle &\mapsto |\uparrow\uparrow\rangle \\ |1,0\rangle &\mapsto \tfrac{1}{\sqrt{2}}(|\uparrow\downarrow\rangle + |\downarrow\uparrow\rangle) \\ |1,-1\rangle &\mapsto |\downarrow\downarrow\rangle \end{aligned} \quad (6.5)$$

and the singlet state corresponds to,

$$|0,0\rangle \mapsto \frac{1}{\sqrt{2}}(|\uparrow\downarrow\rangle - |\downarrow\uparrow\rangle), \quad (6.6)$$

where upon designation of the top assignment $|1, 1\rangle$, each of the other states are obtained by application of the ladder operators defined in Equation (5.6). For example,

$$S_-|1,1\rangle = \hbar\sqrt{2}|1,0\rangle \quad (6.7)$$

To change to a representation in the uncoupled basis, we write,

$$|1,0\rangle = \frac{1}{\hbar\sqrt{2}} S_-|1,1\rangle = \frac{1}{\hbar\sqrt{2}}(S_{1,-} + S_{2,-})|\uparrow\uparrow\rangle, \quad (6.8)$$

and therefore,

$$|1,0\rangle = \frac{1}{\hbar\sqrt{2}}(S_{1,-}|\uparrow\uparrow\rangle + S_{2,-}|\uparrow\uparrow\rangle), \quad (6.9)$$

which yields,

$$|1,0\rangle = \frac{1}{\hbar\sqrt{2}}(|\downarrow\uparrow\rangle + |\uparrow\downarrow\rangle), \quad (6.10)$$

Finally to obtain the singlet state, we simply orthogonalize the above, yielding,

$$|0,0\rangle = \frac{1}{\hbar\sqrt{2}}(|\downarrow\uparrow\rangle - |\uparrow\downarrow\rangle) \quad (6.11)$$

The above procedure is easily carried out by a computer for any combination of particle spins. For many particle systems, for instance in the computation of nuclear spins or non-ground state

^{1}H configurations, the associativity property is used and the above description applies. In the above example, the factors $\frac{1}{\hbar\sqrt{2}}$ are the Clebsch-Gordan coefficients. There are several openly available implementations of Clebsch-Gordan coefficient calculators, in addition to tabulations in handbooks of physics formulae [119, 1].

Chapter 7

Group Theory: SO(3), SU(2), and SU(3)

I have long held an opinion, almost amounting to conviction, in common I believe with many other lovers of natural knowledge, that the various forms under which the forces of matter are made manifest have one common origin; or, in other words, are so directly related and mutually dependent, that they are convertible, as it were, one into another, and possess equivalents of power in their action. [*]

—Michael Faraday

---

[*]'Action of magnets on light' In: 'On the magnetization of light and the illumination of magnetic lines of force'. In: Experimental Researches in Electricity–Nineteenth series. In: *Philosophical Transactions of the Royal Society of London* (1846) vol. 136 chapt. 26 sect. 2146.

76 CHAPTER 7. GROUP THEORY: SO(3), SU(2), AND SU(3)

Orbital angular momentum and spin are the source of magnetic resonance. Their abstract mathematical description is group theoretic, and extends naturally into other fundamental physics such as the strong interaction of quarks in nucleons and nucleons in nuclei. The Frenchman Henri Poincaré commented that "mathematicians do not study objects, but the relationships between objects". Relationships between symmetry groups have indeed been the essential device for probing the unseen in the realm of high energy particle physics.

For spin 1/2 particles such as the electron and the quark, the spin observable can be represented by the SU(2) group. SU(N) is the group of $N \times N$ unitary matrices with unit determinant, in which the group operation is matrix-matrix multiplication. Similarly SO(N) is the group of $N \times N$ orthogonal matrices of unit determinant, with matrix multiplication as group operator. Analogous to the SU(2) representation of spin, quantum orbital angular momentum can be represented by the SO(3) group. Both groups share a Lie Algebra, encoded in the commutation relations shown above, because they are isomorphic to each other. In particular, SU(2) is a double cover of SO(3). The SU(3) group models the interactors of the strong force in Quantum Chromodynamics (QCD).

7.1 SO(3)

A counterclockwise rotation by an angle ϕ about the x, y, or z-axes, can respectively be represented by

$$R_x(\phi) = \begin{pmatrix} 1 & 0 & 0 \\ 0 & \cos\phi & -\sin\phi \\ 0 & \sin\phi & \cos\phi \end{pmatrix},$$

$$R_y(\phi) = \begin{pmatrix} \cos\phi & 0 & \sin\phi \\ 0 & 1 & 0 \\ -\sin\phi & 0 & \cos\phi \end{pmatrix}, \qquad (7.1)$$

$$R_z(\phi) = \begin{pmatrix} \cos\phi & -\sin\phi & 0 \\ \sin\phi & \cos\phi & 0 \\ 0 & 0 & 1 \end{pmatrix}.$$

The corresponding infinitesimal generators, G_x, G_y, G_z respectively are:

7.2. SU(2)

$$\begin{pmatrix} 0 & 0 & 0 \\ 0 & 0 & -1 \\ 0 & 1 & 0 \end{pmatrix}, \begin{pmatrix} 0 & 0 & 1 \\ 0 & 0 & 0 \\ -1 & 0 & 0 \end{pmatrix}, \begin{pmatrix} 0 & -1 & 0 \\ 1 & 0 & 0 \\ 0 & 0 & 0 \end{pmatrix}. \quad (7.2)$$

The corresponding Lie Algebra can then be readily shown to be:

$$[G_i, G_j] = i\hbar \epsilon_{ijk} G_k \quad (7.3)$$

7.2 SU(2)

$$SU(2) = \left\{ \begin{pmatrix} a & -b^* \\ b & a^* \end{pmatrix} : a, b \in \mathbf{C}, a^2 + b^2 = 1 \right\} \quad (7.4)$$

It follows that the generators of SU(2) are the Pauli Matrices given by,

$$\sigma_1 = \begin{pmatrix} 0 & 1 \\ 1 & 0 \end{pmatrix}, \quad \sigma_2 = \begin{pmatrix} 0 & -i \\ i & 0 \end{pmatrix}, \quad \sigma_3 = \begin{pmatrix} 1 & 0 \\ 0 & -1 \end{pmatrix},$$
$$(7.5)$$

The SU(2) Lie Algebra is then prescribed by the following commutation and anti-commutation relations,

$$[\sigma_a, \sigma_b] = 2i\epsilon_{abc}\sigma_c, \quad (7.6)$$

and

$$\{\sigma_a, \sigma_b\} = 2\delta_{ab} \cdot I + i\epsilon_{abc}\sigma_c, \quad (7.7)$$

which combine to give:

$$\sigma_a \sigma_b = \delta_{ab} \cdot I + i\epsilon_{abc}\sigma_c, \quad (7.8)$$

where δ_{ab} is the Kronecker delta and $\{x, y\} := xy + yx$ is the anti-commutator.

7.3 SU(2) is Isomorphic to the 3-Sphere

Given $U \in SU(2)$, such that,

$$U = \begin{pmatrix} a & -b^* \\ b & a^* \end{pmatrix},$$

We define an assignment, ζ, such that

$$\zeta : \begin{cases} a \mapsto x_0 + ix_3 \\ b \mapsto -x_2 + ix_1 \end{cases}, \quad (7.9)$$

where $x_0, x_1, x_2, x_3 \in \Re$. It follows that,

$$U = x_0 I + i x_p \sigma_p, \quad (7.10)$$

where I is the 2×2 identity matrix, σ_p are the Pauli matrices, and we have used the Einstein summation notation over p.

$$det(U) = 1 \Rightarrow a^2 + b^2 = 1 \Rightarrow x_0^2 + x_p x_p = 1 \Rightarrow \mathbf{x} \in S^3 \quad (7.11)$$

ζ is a bijective homomorphism between $SU(2)$ and S^3. And this shows $SU(2)$ is isomorphic to S^3.

7.4 SU(3)

The generators of the SU(3) group are given by the Gell-Mann matrices:

$$\lambda_1 = \begin{pmatrix} 0 & 1 & 0 \\ 1 & 0 & 0 \\ 0 & 0 & 0 \end{pmatrix}, \quad \lambda_2 = \begin{pmatrix} 0 & -i & 0 \\ i & 0 & 0 \\ 0 & 0 & 0 \end{pmatrix},$$

$$\lambda_3 = \begin{pmatrix} 1 & 0 & 0 \\ 0 & -1 & 0 \\ 0 & 0 & 0 \end{pmatrix}, \quad \lambda_4 = \begin{pmatrix} 0 & 0 & 1 \\ 0 & 0 & 0 \\ 1 & 0 & 0 \end{pmatrix},$$

$$\lambda_5 = \begin{pmatrix} 0 & 0 & -i \\ 0 & 0 & 0 \\ i & 0 & 0 \end{pmatrix}, \quad \lambda_6 = \begin{pmatrix} 0 & 0 & 0 \\ 0 & 0 & 1 \\ 0 & 1 & 0 \end{pmatrix},$$

$$\lambda_7 = \begin{pmatrix} 0 & 0 & 0 \\ 0 & 0 & -i \\ 0 & i & 0 \end{pmatrix}, \quad \lambda_8 = \frac{1}{\sqrt{3}} \begin{pmatrix} 1 & 0 & 0 \\ 0 & 1 & 0 \\ 0 & 0 & -2 \end{pmatrix},$$

(7.12)

Then defining $T_j = \frac{\lambda_j}{2}$, the commutation relations are,

$$[T_a, T_b] = i \sum_{c=1}^{8} f_{abc} T_c, \quad (7.13)$$

7.4. SU(3)

and

$$\{T_a, T_b\} = \frac{4}{3}\delta_{ab} + 2\sum_{c=1}^{8} d_{abc}T_c, \qquad (7.14)$$

where f_{abc} and d_{abc} are the antisymmetric and symmetric structure constants of SU(3) respectively.

Chapter 8

MRI Signals Processing

But still try, for who knows what is possible...*

—Michael Faraday

**The Life and Letters of Faraday* (1870), H. B. Jones (Ed.), Vol. 2, p.483

Signals processing is playing an increasingly important role in magnetic resonance imaging and related applications. Some of the factors driving this trend include:

1. The free inductance decay signal is often collected in the frequency domain and Fourier transformed into the time domain [38].

2. In correspondence with the gradual shift towards MRI scanners of higher magnetic field, the signal-to-noise ratio requires increasingly more effective noise filtering algorithms.

3. As signal generation and collection methods become increasingly more sophisticated e.g via spirals and other geometric sequences, non-uniform sampling, and exotic B_1 pulse sequence configurations, signals processing methods have needed to advance in tandem to address problems arising.

4. Imaging and graphics applications have significantly driven the interest and funding for fast algorithms and computing. Notable advancements include the fast Fourier transform, the fast-multipole method of Greengard and Rokhlin [44, 45], and parallel computing hardware such as the Compute Unified Device Architecture (CUDA) multi-core processors by Nvidia for graphics processing [90].

5. A new field is taking form at the interface of high performance software algorithms and multi-core hardware processors [88, 69, 87, 49, 91, 55].

We take a brief look at some of the basics of signals processing below,

8.1 Continuous and Discrete Fourier Transforms

The Fourier transform converts a time domain signal to its corresponding frequency domain signal, and is given by,

$$F[f(t)] = f(\omega) = \int_{-\infty}^{\infty} f(t)e^{-i\omega t}dt, \qquad (8.1)$$

while the inverse Fourier transform converts a frequency domain signal to its corresponding time domain signal, and is given by,

$$F^{-1}[f(\omega)] = f(t) = \int_{-\infty}^{\infty} f(\omega)e^{i\omega t}d\omega. \quad (8.2)$$

In practice, the discrete versions of the above equations are used instead, and give good results provided that signal data is sampled at an appropriate frequency, the Nyquist frequency. Given an MRI signal of N sample points, $f[n]$, where $n = 0, .., N-1$, the frequency domain representation is given by the discrete Fourier transform,

$$f[\omega_k] = \sum_{n=0}^{N-1} f[n]e^{-i2\pi \frac{k}{N} n}, \quad (8.3)$$

and the discrete inverse Fourier transform is given by,

$$f[n] = \frac{1}{N} \sum_{k=0}^{N-1} f[\omega_k]e^{i2\pi \frac{k}{N} n}. \quad (8.4)$$

The above Fourier formulas readily allow extension to two and three dimensions as well as various other customizations, tailored to suite the particular MRI application and data format.

8.2 MRI Sampling and the Aliasing Problem

The Whittaker- Kotelnikov- Shannon- Raabe- Someya- Nyquist theorem prescribes a lower bound on the sampling frequency necessary for perfect reconstruction of a band-limited signal. It states that given a band limited signal $g(t)$ whose frequency domain representation is such that $|g(f)| < B$ for all f and some B, then by sampling at a rate $f_s = 2B$, the image can be perfectly reconstructed. The sampling interval is $T = 1/f_s$, and the discrete time signal is expressed as $g[nT]$ where n is an integer. $f_s = \omega_s/2\pi$ and is in units of hertz, hence T is in seconds. The perfect reconstruction is given by the Whittaker-Shannon interpolation formula,

$$g(t) = \sum_{n=-\infty}^{\infty} g[nT] \cdot \text{sinc}\left(\frac{t - nT}{T}\right), \quad (8.5)$$

where sinc is the *normalized sampling function* given by,

$$\text{sinc}(x) = \begin{cases} 1 & \text{if } x = 0, \\ \dfrac{\sin(\pi x)}{\pi x} & \text{otherwise,} \end{cases} \quad (8.6)$$

84 CHAPTER 8. MRI SIGNALS PROCESSING

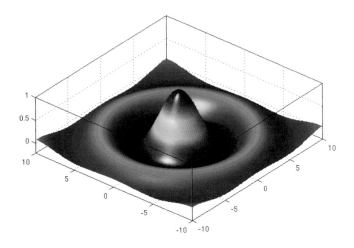

Figure 8.1: 3D Sinc Function

and is shown in 3-dimenisonal form in Figure (8.1). In practice, low-pass filtering methods can be used to decrease the amplitude of frequency components which exceed the effective bandlimit.

The aliasing problem arises when the sampling rate f_s is less than twice the bandlimit B. The effect of this is that the reconstructed image is an alias of the actual image. The Discrete Time Fourier Transform (DTFT) of a signal is given by,

$$X(\omega) = \sum_{n=-\infty}^{\infty} x[n]e^{-i\omega n}, \qquad (8.7)$$

and in the case of an under-sampled MRI signal, the DTFT matches that of the alias. Therefore the reconstructed image is merely an alias. This phenomenon can be described more formally by invoking the poisson summation formula,

$$X_s(f) := \sum_{-\infty}^{\infty} X(f - kf_s) = \sum_{-\infty}^{\infty} T \cdot x(nT)e^{-i2\pi fnT}, \qquad (8.8)$$

where the middle term above is the periodic summation. For a bandlimited under-sampled signal, increasing the sampling rate to satisfy the Nyquist criterion will restore perfect reconstruction.

8.3 The Convolution Theorem

The convolution theorem facilitates application-specific image processing before or after reconstruction in either k-space or time domain. The theorem is given by,

$$F\{(f * g)(t)\} = f(\omega) \cdot g(\omega), \tag{8.9}$$

and states that the Fourier transform of the convolution of two functions equals the product of the Fourier transforms of both functions. Where the convolution is defined as,

$$(f * g)(t) := \int_{-\infty}^{\infty} f(\tau) g(t - \tau) d\tau. \tag{8.10}$$

Chapter 9

Future Research Directions

Nothing is too wonderful to be true, if it be consistent with the laws of nature; and ...experiment is the best test of such consistency.[*]

—Michael Faraday

---

[*]Labratory journal entry #10,040, March 19th 1849; from: *The Life and Letters of Faraday* (1870), H. B. Jones (ed.) Vol. 2, p. 253.

88   CHAPTER 9. FUTURE RESEARCH DIRECTIONS

Upon acquiring a grasp of quantum mechanics and solidifying that grasp by a real world application such as the MRI, the curious reader may begin to wander what new applications of quantum mechanics are on the horizon both in medicine and in other fields. This final chapter provides a glimpse of one of the exciting sides of the applied physics frontier. In what follows, we briefly overview quantum information theory, quantum entanglement, quantum computing, quantum cyber security, and quantum optics.

9.1 Quantum Information Theory

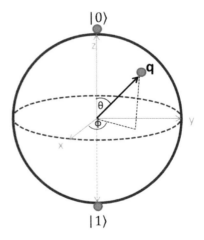

Figure 9.1: Bloch Sphere

Quantum information theory is the quantum mechanical analogue of classical information theory. While classical information theory is based on the *bit* as the unit of information, quantum information theory is based on the *qubit*, defined below.

The Qubit

The qubit is the quantum mechanical analogue of the bit. Classically the bit takes on values 0 (off state, $|0\rangle$) or 1 (on state, $|1\rangle$). An important distinction is the qubit, q, can be in the on state, the off state, or any quantum superposition of on and off states. This is represented by,

$$q = \alpha|0\rangle + \beta|1\rangle, \qquad (9.1)$$

where the probability amplitudes α and β are complex number for which

$$|\alpha|^2 + |\beta|^2 = 1. \qquad (9.2)$$

The Bloch Sphere

The Bloch sphere is a plot of all possible values of the qubit vector. Figure (9.1) depicts the Bloch sphere. The polar angle θ encodes the probability amplitudes, α and β. The azimuthal angle ϕ on the other hand, has no physical meaning in most current interpretations of quantum information theory. Unitary transformations do not change the qubit magnitude $|q|$ and therefore correspond to rotations on the Bloch sphere.

Collapse of the Wave Function

Upon measurement of a qubit q, the obtained value is due to either $|0\rangle$ or $|1\rangle$, but not both. Specifically, with a probability $|\alpha|^2$, the wave function collapses into the state $|0\rangle$ and with a probability $|\beta|^2$ it collapses into the state $|1\rangle$. This view of measurement originated with the Copenhagen interpretation of quantum mechanics.

9.2 Quantum Entanglement

Quantum entanglement is a phenomenon of strongly correlated quantum mechanical systems, in which upon separation, measurement of one part of the system determines the state of the other part. Take for instance the following superimposition of two two-particle systems known as the Bell state,

$$\psi = \frac{1}{\sqrt{2}}\left(|00\rangle + |11\rangle\right). \qquad (9.3)$$

In the uncoupled basis, say we give the first particle to Alice who resides in Alpha Centauri and the second particle to Bob who resides in Crab nebula. If Alice makes a measurement and obtains a value $|0\rangle$ in Alpha Centauri, then Bob necessarily also gets a value $|0\rangle$ in Crab nebula, because the only possible states in the coupled basis are $|00\rangle$ and $|11\rangle$. This state change occurs instantaneously despite the stars being very distant from each other. Of note, however, superluminous communication of classical information is nonetheless precluded.

The study of quantum entanglement began with Albert Einstein, Boris Podolsky, and Nathan Rosen's early criticism of quantum mechanics as incomplete due its allowance of a phenomenon as strange as quantum entanglement, which was later referred to by Einstein as "spooky action-at-a-distance". In 1935 they published an influential paper presenting what is now called the Einstein-Podolsky-Rosen (EPR) paradox [37]. Einstein's still incompletely resolved concern was the apparent violation of special relativity, by superluminous transmission of information through instantaneous non-local state change. Though certainly still controversial, there is now some growing evidence in support of quantum entanglement. Furthermore, there are ongoing attempts to harness it for quantum communication and quantum computing.

9.3 Quantum Computing

Quantum computing is the idea of using the qubit to store information and quantum entanglement (or other quantum mechanical correlation) to compute and transmit information. Quantum computing is promoted by its proponents as both feasible and having the potential to enormously increase computation and data storage capacity compared to classical computers. A single qubit can theoretically encode much more information, compared to a classical bit which can only deterministically take on values of 0 or 1. In reality however, the physical systems being investigated for qubit implementation, naturally place physical constraints on the degrees of freedom. Nonetheless, the possibilities are as exciting as the abundance of unanswered questions in this active area of research. Below we look at some potential ways of implementing the qubit.

Physical Realization

Various candidate physical systems have been proposed and are being investigated for qubit implementation and storage. A sampling of some of the more widely studied systems include,

The Electron

The electron is a spin 1/2 particle which can be either in the spin up state, $|\uparrow\rangle$, the spin down state, $|\downarrow\rangle$, or some quantum superposition of the two, $\alpha|\uparrow\rangle + \beta|\downarrow\rangle$.

The Nucleus

As discussed above, the nucleus also has spin and is a possible hardware implementation of the qubit. Nuclear magnetic resonance pulses can conceivably be used as switches into the hardware circuit.

The Photon

The photon is a spin 1 particle or vector boson which can be used as qubit implementation in a number of ways including,

- Polarization state: Horizontal vs. Vertical
- Fock state: Vacuum state vs. Occupancy number n=1
- Orbital angular momentum state [63, 116, 2, 74, 80]: Clockwise vs. Counterclockwise

Other Qubit Implementation Candidates

Other potential hardware implementations of the qubit under investigation include squeezed coherent states of light, Josephson junctions in superconductivity, optical lattices, and quantum dots.

9.4 Quantum Cyber Security

Quantum cyber security or quantum cryptography is the use of quantum communication and quantum computation in cryptography. It exploits wave function collapse, such that attempts to access preveliged data results in a change of the state of the data.

92 CHAPTER 9. FUTURE RESEARCH DIRECTIONS

This simultaneously corrupts the attackers stolen data, and notifies the owner of the breach. The feasibility and timeline of practical applications of quantum computing and cryptography are currently unclear. Any substantial realization of quantum computing will immediately render classically encrypted keys insecure. Hence quantum cryptography must be developed in tandem.

9.5 Quantum Optics

Quantum optics was not only present at the birth of quantum mechanics, it was simultaneously born of the same mother, Max Planck's attempts to understand Black-body radiation. Though quantum optics remains hardly separable from areas like atomic physics, solid state physics, and condensed matter physics, it has gradually grown into its own unique identity. Subfields identified under quantum optics include Laser science, photonics in a sense analogous to electronics, and coherent states such as Bose-Einstein condensation. One attractive feature of quantum optics is its seamless interface between application and theory. A few exciting applications of quantum optics in active development include: Orbital angular momentum of light for rotation of matter [63, 116, 2, 74, 80], parametric down conversion for single photon production, and optical tweezers for planar translocation of matter with laser beams.

Bibliography

[1] M. Abramowitz and I. A. Stegun, *Handbook of mathematical functions: with formulas, graphs, and mathematical tables*, vol. 55, Courier Dover Publications, 1964.

[2] L. Allen, M. W. Beijersbergen, R. J. C. Spreeuw, and J. P. Woerdman, *Orbital angular momentum of light and the transformation of laguerre-gaussian laser modes*, Phys. Rev. A **45** (1992), 8185–8189.

[3] H. Ay, F. S. Buonanno, G. Rordorf, P. W. Schaefer, L. H. Schwamm, O. Wu, R. G. Gonzalez, K. Yamada, G. A. Sorensen, and W. J. Koroshetz, *Normal diffusion-weighted MRI during stroke-like deficits*, Neurology **52** (1999), no. 9, 1784–1784.

[4] D. Azzopardi, J. S. Wyatt, E. B. Cady, D. T. Delpy, J. Baudin, A. L. Stewart, P. L. Hope, P. A. Hamilton, and E. O. R. Reynolds, *Prognosis of newborn infants with hypoxic-ischemic brain injury assessed by phosphorus magnetic resonance spectroscopy*, Pediatric Research **25** (1989), no. 5, 445–451.

[5] M. Babic, D. Horak, P. Jendelova, V. Herynek, V. Proks, V. Vanecek, P. Lesny, and E. Sykova, *The use of dopamine-hyaluronate associate-coated maghemite nanoparticles to label cells*, International Journal of Nanomedicine **7** (2012), 1461.

[6] P. A. Bandettini and E. C. Wong, *A hypercapnia-based normalization method for improved spatial localization of human brain activation with fMRI*, NMR in Biomedicine **10** (1997), no. 4-5, 197–203.

[7] P. A. Barber, D. G. Darby, P. M. Desmond, Q. Yang, R. P. Gerraty, D. Jolley, G. A. Donnan, B. M. Tress, and S. M. Davis, *Prediction of stroke outcome with echoplanar perfusion-and diffusion-weighted MRI*, Neurology **51** (1998), no. 2, 418–426.

[8] J. Bardeen, *Theory of the meissner effect in superconductors*, Physical Review **97** (1955), no. 6, 1724.

[9] J. Bardeen, L. N. Cooper, and J. R. Schrieffer, *Microscopic theory of superconductivity*, Physical Review **106** (1957), no. 1, 162–164.

[10] ———, *Theory of superconductivity*, Physical Review **108** (1957), no. 5, 1175.

[11] A. J. Barkovich, K. D. Westmark, H. S. Bedi, J. C. Partridge, D. M. Ferriero, and D. B. Vigneron, *Proton spectroscopy and diffusion imaging on the first day of life after perinatal asphyxia: preliminary report*, American Journal of Neuroradiology **22** (2001), no. 9, 1786–1794.

[12] P. Barzó, A. Marmarou, P. Fatouros, K. Hayasaki, and F. Corwin, *Contribution of vasogenic and cellular edema to traumatic brain swelling measured by diffusion-weighted imaging*, Journal of Neurosurgery **87** (1997), no. 6, 900–907.

[13] C. L. Bennett, Z. P. Qureshi, A. O. Sartor, L. A. B. Norris, A. Murday, S. Xirasagar, and H.S . Thomsen, *Gadolinium-induced nephrogenic systemic fibrosis: the rise and fall of an iatrogenic disease*, Clinical Kidney Journal **5** (2012), no. 1, 82–88.

[14] M. A. Bernstein, K. F. King, and X. J. Zhou, *Handbook of MRI pulse sequences*, Academic Press, 2004.

93

[15] M. N. Biltcliffe, P. E. Hanley, J. B. McKinnon, and P. Roubeau, *The operation of superconducting magnets at temperatures below 4.2 k*, Cryogenics **12** (1972), no. 1, 44–47.

[16] F. Binkofski and R. J. Seitz, *Modulation of the BOLD-response in early recovery from sensorimotor stroke*, Neurology **63** (2004), no. 7, 1223–1229.

[17] E. Brücher, *Kinetic stabilities of gadolinium (III) chelates used as MRI contrast agents*, Contrast Agents I, Topics in Current Chemistry (2002), **221** 103–122.

[18] J. W. M. Bulte and D. L. Kraitchman, *Iron oxide MR contrast agents for molecular and cellular imaging*, NMR in Biomedicine **17** (2004), no. 7, 484–499.

[19] K. Butcher, M. Parsons, T. Baird, A. Barber, G. Donnan, P. Desmond, B. Tress, and S. Davis, *Perfusion thresholds in acute stroke thrombolysis*, Stroke **34** (2003), no. 9, 2159–2164.

[20] C. Calautti and J. C. Baron, *Functional neuroimaging studies of motor recovery after stroke in adults*, Stroke **34** (2003), no. 6, 1553–1566.

[21] M. Cappellini, G. Rapisardi, M. L. Cioni, and C. Fonda, *Acute hypoxic encephalopathy in the full-term newborn: correlation between magnetic resonance spectroscopy and neurological evaluation at short and long term.*, La Radiologia Medica **104** (2002), no. 4, 332.

[22] P. Caravan, *Strategies for increasing the sensitivity of gadolinium based MRI contrast agents*, Chem. Soc. Rev. **35** (2006), no. 6, 512–523.

[23] _____, *Protein-targeted gadolinium-based magnetic resonance imaging (MRI) contrast agents: design and mechanism of action*, Accounts of Chemical Research **42** (2009), no. 7, 851–862.

[24] P. Caravan, J. J. Ellison, T. J. McMurry, and R. B. Lauffer, *Gadolinium (III) chelates as MRI contrast agents: structure, dynamics, and applications*, Chemical Reviews **99** (1999), no. 9, 2293–2352.

[25] S. C. Chang, P. H. Lai, W. L. Chen, H. H. Weng, J. T. Ho, J. S. Wang, C. Y. Chang, H. B. Pan, and C. F. Yang, *Diffusion-weighted MRI features of brain abscess and cystic or necrotic brain tumors: comparison with conventional MRI*, Clinical Imaging **26** (2002), no. 4, 227–236.

[26] L. N. Cooper, *Bound electron pairs in a degenerate Fermi gas*, Physical Review **104** (1956), no. 4, 1189.

[27] S. B. Coutts, J. E. Simon, A. I. Tomanek, P. A. Barber, J. Chan, M. E. Hudon, J. R. Mitchell, R. Frayne, M. Eliasziw, A. M. Buchan, and A. M. Demchuk, *Reliability of assessing percentage of diffusion-perfusion mismatch*, Stroke **34** (2003), no. 7, 1681–1683.

[28] S. C. Cramer, R. Shah, J. Juranek, K. R. Crafton, and V. Le, *Activity in the peri-infarct rim in relation to recovery from stroke*, Stroke **37** (2006), no. 1, 111–115.

[29] R. A. Crisostomo, M. M. Garcia, and D. C. Tong, *Detection of diffusion-weighted MRI abnormalities in patients with transient ischemic attack*, Stroke **34** (2003), no. 4, 932–937.

[30] E. J. Cukauskas, D. A. Vincent, and B. S. Deaver, *Magnetic susceptibility measurements using a superconducting magnetometer*, Review of Scientific Instruments **45** (1974), no. 1, 1–6.

[31] R. Damadian, *Tumor detection by nuclear magnetic resonance*, Science **171** (1971), no. 3976, 1151–1153.

[32] R. Damadian, K. Zaner, D. Hor, T. DiMaio, L. Minkoff, and M. Goldsmith, *Nuclear magnetic resonance as a new tool in cancer research: human tumors by nmr*, Annals of the New York Academy of Sciences **222** (1973), no. 1, 1048–1076.

[33] R. V. Damadian, *Apparatus and method for detecting cancer in tissue*, 1974, US Patent 3,789,832.

BIBLIOGRAPHY 95

[34] P.A.M. Dirac, *The principles of quantum mechanics*, vol. 27, Oxford University Press, USA, (1982).

[35] C. L. Dumoulin and H. R. Hart Jr, *Magnetic resonance angiography.*, Radiology **161** (1986), no. 3, 717–720.

[36] C. L. Dumoulin, S. P. Souza, M. F. Walker, and W. Wagle, *Three-dimensional phase contrast angiography*, Magnetic Resonance in Medicine **9** (1989), no. 1, 139–149.

[37] A. Einstein, B. Podolsky, and N. Rosen, *Can quantum-mechanical description of physical reality be considered complete?*, Physical Review **47** (1935), no. 10, 777.

[38] R. R. Ernst and W. A. Anderson, *Application of fourier transform spectroscopy to magnetic resonance*, Review of Scientific Instruments **37** (1966), no. 1, 93–102.

[39] J. Faiz Kayyem, R. M. Kumar, S. E. Fraser, and T. J. Meade, *Receptor-targeted co-transport of DNA and magnetic resonance contrast agents*, Chemistry & biology **2** (1995), no. 9, 615–620.

[40] D. N. Firmin, G. L. Nayler, P. J. Kilner, and D. B. Longmore, *The application of phase shifts in NMR for flow measurement*, Magnetic Resonance in Medicine **14** (1990), no. 2, 230–241.

[41] B. S. Geisler, F. Brandhoff, J. Fiehler, C. Saager, O. Speck, J. Röther, H. Zeumer, and T. Kucinski, *Blood oxygen level-dependent MRI allows metabolic description of tissue at risk in acute stroke patients*, Stroke **37** (2006), no. 7, 1778–1784.

[42] C. B. Grandin, T. P. Duprez, A. M. Smith, C. Oppenheim, A. Peeters, A. R. Robert, and G. Cosnard, *Which MR-derived perfusion parameters are the best predictors of infarct growth in hyperacute stroke? comparative study between relative and quantitative measurements*, Radiology **223** (2002), no. 2, 361–370.

[43] D. J. Griffiths, *Introduction to quantum mechanics*, Pearson Prentice Hall, Upper Saddle River, NJ, (2005).

[44] L. Greengard and V. Rokhlin, *A fast algorithm for particle simulations*, Journal of Computational Physics **73** (1987), no. 2, 325–348.

[45] ———, *A new version of the fast multipole method for the laplace equation in three dimensions*, Acta Numerica **6** (1997), no. 1, 229–269.

[46] O. H. J. Gröhn and R. A. Kauppinen, *Assessment of brain tissue viability in acute ischemic stroke by BOLD MRI*, NMR in Biomedicine **14** (2001), no. 7-8, 432–440.

[47] F. Hamzei, R. Knab, C. Weiller, and J. Röther, *The influence of extra-and intracranial artery disease on the BOLD signal in FMRI*, Neuroimage **20** (2003), no. 2, 1393–1399.

[48] P. Hanover and B. Eilhardt, *Conductor system for superconducting cables*, January 1 1972, US Patent 3,634,597.

[49] P. Harish and P. Narayanan, *Accelerating large graph algorithms on the GPU using CUDA*, High Performance Computing–HiPC 2007 (2007), 197–208.

[50] M. Hermier and N. Nighoghossian, *Contribution of susceptibility-weighted imaging to acute stroke assessment*, Stroke **35** (2004), no. 8, 1989–1994.

[51] M. L. Hugh, *Low temperature electric transmission systems*, 1971, US Patent 3,562,401.

[52] G. E. Jackson, S. Wynchank, and M. Woudenberg, *Gadolinium (III) complex equilibria: the implications for Gd (III) MRI contrast agents*, Magnetic Resonance in Medicine **16** (1990), no. 1, 57–66.

[53] C. W. Jung and P. Jacobs, *Physical and chemical properties of superparamagnetic iron oxide MR contrast agents: ferumoxides, ferumoxtran, feruxmoxsil*, Magnetic Resonance Imaging **13** (1995), no. 5, 661–674.

[54] G. W. Kabalka, E. Buonocore, K. Hubner, M. Davis, and L. Huang, *Gadolinium-labeled liposomes containing paramagnetic amphipathic agents: Targeted MRI contrast agents for the liver*, Magnetic Resonance in Medicine **8** (1988), no. 1, 89–95.

[55] S. Kestur, K. Irick, S. Park, A. Al Maashri, V. Narayanan, and C. Chakrabarti, *An algorithm-architecture co-design framework for gridding reconstruction using FPGAs*, Design Automation Conference (DAC), 2011 48th ACM/EDAC/IEEE, IEEE, 2011, pp. 585–590.

[56] Y. R. Kim, I. J. Huang, S. R. Lee, E. Tejima, J. B. Mandeville, M. P. A. van Meer, G. Dai, Y. W. Choi, R. M. Dijkhuizen, E. H. Lo, and B. R. Rosen, *Measurements of BOLD/CBV ratio show altered fMRI hemodynamics during stroke recovery in rats*, Journal of Cerebral Blood Flow & Metabolism **25** (2005), no. 7, 820–829.

[57] H. Kobayashi, M. W. Brechbiel, *Dendrimer-based macromolecular MRI contrast agents: characteristics and application*, Molecular Imaging **2** (2003), no. 1, 1.

[58] S. D. Konda, M. Aref, S. Wang, M. Brechbiel, and E. C. Wiener, *Specific targeting of folate-dendrimer MRI contrast agents to the high affinity folate receptor expressed in ovarian tumor xenografts*, Magnetic Resonance Materials in Physics, Biology and Medicine **12** (2001), no. 2, 104–113.

[59] P. H. Kuo, E. Kanal, A. K. Abu-Alfa, and S. E. Cowper, *Gadolinium-based MR contrast agents and nephrogenic systemic fibrosis*, Radiology **242** (2007), no. 3, 647–649.

[60] S. M. Lai, T. Y. Tsai, C. Y. Hsu, J. L. Tsai, M. Y. Liao, and P. S. Lai, *Bifunctional silica-coated superparamagnetic FePt nanoparticles for fluorescence/MR dual imaging*, Journal of Nanomaterials **2012** (2012), 5.

[61] P. C. Lauterbur, *Image formation by induced local interactions: examples employing nuclear magnetic resonance*, Nature **242** (1973), no. 5394, 190–191.

[62] N. Lee, Y. Choi, Y. Lee, M. Park, W. K. Moon, S. H. Choi, and T. Hyeon, *Water-dispersible ferrimagnetic iron oxide nanocubes with extremely high r 2 relaxivity for highly sensitive in vivo MRI of tumors*, Nano Letters **12** (2012), no. 6, 3127–3131.

[63] J. Leach, M. J. Padgett, S. M. Barnett, S. Franke-Arnold, and J. Courtial, *Measuring the orbital angular momentum of a single photon*, Phys. Rev. Lett. **88** (2002), 257901.

[64] D. R. Lide, *CRC handbook of chemistry and physics*, CRC press, 2012.

[65] M. J. Lipinski, V. Amirbekian, J. C. Frias, J. G. S. Aguinaldo, V. Mani, K. C. Briley-Saebo, J. T. Fallon, E. A. Fisher, and Z. A. Fayad, *MRI to detect atherosclerosis with gadolinium-containing immunomicelles targeting the macrophage scavenger receptor*, Magnetic Resonance in Medicine **56** (2006), no. 3, 601–610.

[66] A. Y. Liu, J. A. Maldjian, L. J. Bagley, G. P. Sinson, and R. I. Grossman, *Traumatic brain injury: diffusion-weighted MR imaging findings*, American Journal of Neuroradiology **20** (1999), no. 9, 1636–1641.

[67] D. H. Live and S. I. Chan, *Bulk susceptibility corrections in nuclear magnetic resonance experiments using superconducting solenoids*, Analytical Chemistry **42** (1970), no. 7, 791–792.

[68] G. K. Malik, M. Pandey, R. Kumar, S. Chawla, B. Rathi, and R. K. Gupta, *MR imaging and in vivo proton spectroscopy of the brain in neonates with hypoxic ischemic encephalopathy*, European Journal of Radiology **43** (2002), no. 1, 6–13.

[69] S.A. Manavski, *CUDA compatible GPU as an efficient hardware accelerator for AES cryptography*, Signal Processing and Communications, 2007. ICSPC 2007. IEEE International Conference on, IEEE, 2007, pp. 65–68.

[70] D. M. Mandell, J. S. Han, J. Poublanc, A. P. Crawley, J. A. Stainsby, J. A. Fisher, and D. J. Mikulis, *Mapping cerebrovascular reactivity using blood oxygen level-dependent MRI in patients with arterial steno-occlusive disease*, Stroke **39** (2008), no. 7, 2021–2028.

[71] C. Maneru, C. Junque, M. Bargallo, M. Olondo, F. Botet, M. Tallada, J. Guardia, and J. M. Mercader, *1H-MR spectroscopy is sensitive to subtle effects of perinatal asphyxia*, Neurology **57** (2001), no. 6, 1115–1118.

[72] P. Marckmann, L. Skov, K. Rossen, A. Dupont, M. B. Damholt, J. G. Heaf, and H. S. Thomsen, *Nephrogenic systemic fibrosis: suspected causative role of gadodiamide used for contrast-enhanced magnetic resonance imaging*, Journal of the American Society of Nephrology **17** (2006), no. 9, 2359–2362.

[73] M. P. Marks, D. C. Tong, C. Beaulieu, G. W. Albers, A. De Crespigny, and M. E. Moseley, *Evaluation of early reperfusion and IV tPA therapy using diffusion- and perfusion-weighted MRI*, Neurology **52** (1999), no. 9, 1792–1792.

[74] A. Mair, A. Vaziri, G. Weihs, and A. Zeilinger, *Entanglement of the orbital angular momentum states of photons*, Nature **412** (2001), no. 6844, 313–316.

[75] H. T. Mebrahtu, I. V. Borzenets, D. E. Liu, H. Zheng, Y. V. Bomze, A. I. Smirnov, H. U. Baranger, and G. Finkelstein, *Quantum phase transition in a resonant level coupled to interacting leads*, Nature **488** (2012), no. 7409, 61–64.

[76] R. W. Meyerhoff, *Superconducting power transmission*, Cryogenics **11** (1971), no. 2, 91–101.

[77] M. Mikawa, H. Kato, M. Okumura, M. Narazaki, Y. Kanazawa, N. Miwa, and H. Shinohara, *Paramagnetic water-soluble metallofullerenes having the highest relaxivity for MRI contrast agents*, Bioconjugate chemistry **12** (2001), no. 4, 510–514.

[78] J. Mintorovitch, M. E. Moseley, L. Chileuitt, H. Shimizu, Y. Cohen, and P. R. Weinstein, *Comparison of diffusion- and T2-weighted MRI for the early detection of cerebral ischemia and reperfusion in rats*, Magnetic Resonance in Medicine **18** (1991), no. 1, 39–50.

[79] M. Miyazaki, S. Sugiura, F. Tateishi, H. Wada, Y. Kassai, and H. Abe, *Non-contrast-enhanced MR angiography using 3D ECG-synchronized half-fourier fast spin echo*, Journal of Magnetic Resonance Imaging **12** (2000), no. 5, 776–783.

[80] G. Molina-Terriza, J. P. Torres, and L. Torner, *Management of the angular momentum of light: Preparation of photons in multidimensional vector states of angular momentum*, Phys. Rev. Lett. **88** (2001), 013601.

[81] P. J. Mohr, B. N. Taylor, and D. B. Newell, *CODATA recommended values of the fundamental physical constants: 2010*. 2012, E-print: arXiv. org/abs/1203.5425.

[82] B. Morgan, A. L. Thomas, J. Drevs, J. Hennig, M. Buchert, A. Jivan, M. A. Horsfield, K. Mross, H. A. Ball, L. Lee, et al., *Dynamic contrast-enhanced magnetic resonance imaging as a biomarker for the pharmacological response of PTK787/ZK 222584, an inhibitor of the vascular endothelial growth factor receptor tyrosine kinases, in patients with advanced colorectal cancer and liver metastases: results from two phase I studies*, Journal of Clinical Oncology **21** (2003), no. 21, 3955–3964.

[83] M. E. Moseley, Y. Cohen, J. Mintorovitch, L. Chileuitt, H. Shimizu, J. Kucharczyk, M. F. Wendland, and P. R. Weinstein, *Early detection of regional cerebral ischemia in cats: comparison of diffusion- and T2-weighted MRI and spectroscopy*, Magnetic Resonance in Medicine **14** (1990), no. 2, 330–346.

[84] Y. Murata, K. Sakatani, T. Hoshino, N. Fujiwara, T. Kano, S. Nakamura, and Y. Katayama, *Effects of cerebral ischemia on evoked cerebral blood oxygenation responses and BOLD contrast functional MRI in stroke patients*, Stroke **37** (2006), no. 10, 2514–2520.

[85] T. Neumann-Haefelin, H. J. Wittsack, F. Wenserski, M. Siebler, R. J. Seitz, U. Mödder, and H. J. Freund, *Diffusion- and perfusion-weighted MRI: the DWI/PWI mismatch region in acute stroke*, Stroke **30** (1999), no. 8, 1591–1597.

[86] E. A. Neuwelt, B. E. Hamilton, C. G. Varallyay, W. R. Rooney, R. D. Edelman, P. M. Jacobs, and S. G. Watnick, *Ultrasmall superparamagnetic iron oxides (USPIOs): a future alternative magnetic resonance (MR) contrast agent for patients at risk for nephrogenic systemic fibrosis (NSF)?*, Kidney International **75** (2008), no. 5, 465–474.

[87] J. Nickolls, I. Buck, M. Garland, and K. Skadron, *Scalable parallel programming with CUDA*, Queue **6** (2008), no. 2, 40–53.

[88] A. Nukada, Y. Ogata, T. Endo, and S. Matsuoka, *Bandwidth intensive 3-D FFT kernel for GPUs using CUDA*, High Performance Computing, Networking, Storage and Analysis, 2008. SC 2008. International Conference for, IEEE, 2008, pp. 1–11.

[89] N.D. Opdyke and V. Mejia, *Earth's magnetic field*, Geophysical Monograph Series **145** (2004), 315–320.

[90] L. Pan, L. Gu, and J. Xu, *Implementation of medical image segmentation in CUDA*, Information Technology and Applications in Biomedicine, 2008. ITAB 2008. International Conference on, IEEE, 2008, pp. 82–85.

[91] G. Papamakarios, G. Rizos, N .P. Pitsianis, and X. Sun, *Fast computation of local correlation coefficients on graphics processing units*, Proc. of SPIE Vol, vol. 7444, 2009, pp. 744412–1.

[92] M. W. Parsons, P. A. Barber, J. Chalk, D. G. Darby, S. Rose, P. M. Desmond, R. P. Gerraty, B. M. Tress, P. M. Wright, G. A. Donnan, and S. M. Davis, *Diffusion-and perfusion-weighted MRI response to thrombolysis in stroke*, Annals of Neurology **51** (2002), no. 1, 28–37.

[93] J. E. Perthen, F. Calamante, D. G. Gadian, and A. Connelly, *Is quantification of bolus tracking MRI reliable without deconvolution?*, Magnetic Resonance in Medicine **47** (2002), no. 1, 61–67.

[94] O. A. C. Petroff, J. W. Prichard, K. L. Behar, J. R. Alger, J. A. den Hollander, and R. G. Shulman, *Cerebral intracellular pH by 31P nuclear magnetic resonance spectroscopy*, Neurology **35** (1985), no. 6, 781–781.

[95] J. M. Provenzale, R. D. Tien, G. J. Felsberg, and L. Hacein-Bey, *Spinal dural arteriovenous fistula: demonstration using phase contrast MRA.*, Journal of Computer Assisted Tomography **18** (1994), no. 5, 811.

[96] K. N. Raymond, V. C. Pierre, et al., *Next generation, high relaxivity gadolinium MRI agents*, Bioconjugate chemistry **16** (2005), no. 1, 3–8.

[97] G. Rordorf, W. J. Koroshetz, W. A. Copen, S. C. Cramer, P. W. Schaefer, R. F. Budzik, L. H. Schwamm, F. Buonanno, A. G. Sorensen, and G. Gonzalez, *Regional ischemia and ischemic injury in patients with acute middle cerebral artery stroke as defined by early diffusion-weighted and perfusion-weighted MRI*, Stroke **29** (1998), no. 5, 939–943.

[98] H. J. Rosen, S. E. Petersen, M. R. Linenweber, A. Z. Snyder, D. A. White, L. Chapman, A. W. Dromerick, J. A. Fiez, and M. Corbetta, *Neural correlates of recovery from aphasia after damage to left inferior frontal cortex*, Neurology **55** (2000), no. 12, 1883–1894.

[99] M. Rovaris, A. Gass, R. Bammer, S. J. Hickman, O. Ciccarelli, D. H. Miller, and M. Filippi, *Diffusion MRI in multiple sclerosis*, Neurology **65** (2005), no. 10, 1526–1532.

[100] C. Rydahl, H. S. Thomsen, and P. Marckmann, *High prevalence of nephrogenic systemic fibrosis in chronic renal failure patients exposed to gadodiamide, a gadolinium-containing magnetic resonance contrast agent*, Investigative Radiology **43** (2008), no. 2, 141.

[101] J. J. Sakurai, *Modern quantum mechanics*, Addison-Wesley, Reading, MA.– Edited by Tuan, S. F., (1994).

[102] G. Schlaug, A. Benfield, A. E. Baird, B. Siewert, K. O. Lövblad, R. A. Parker, R. R. Edelman, and S. Warach, *The ischemic penumbra*, Neurology **53** (1999), no. 7, 1528–1528.

[103] L. C. Shih, J. L. Saver, J. R. Alger, S. Starkman, M. C. Leary, F. Vinuela, G. Duckwiler, Y. P. Gobin, R. Jahan, J. P. Villablanca, P. M. Vespa, and C. S. Kidwell, *Perfusion-weighted magnetic resonance imaging thresholds identifying core, irreversibly infarcted tissue*, Stroke **34** (2003), no. 6, 1425–1430.

[104] M. B. Singer, J. Chong, D. Lu, W. J. Schonewille, S. Tuhrim, and S. W. Atlas, *Diffusion-weighted MRI in acute subcortical infarction*, Stroke **29** (1998), no. 1, 133–136.

BIBLIOGRAPHY

[105] M. B. Stearns, *Spin-echo and free-induction-decay measurements in pure Fe and Fe-rich ferromagnetic alloys: Domain-wall dynamics*, Physical Review **162** (1967), no. 2, 496.

[106] G. Strangman, J. P. Culver, J. H. Thompson, and D. A. Boas, *A quantitative comparison of simultaneous BOLD fMRI and NIRS recordings during functional brain activation*, Neuroimage **17** (2002), no. 2, 719–731.

[107] A. Sturzu, S. Sheikh, U. Klose, M. Deeg, H. Echner, T. Nagele, U. Ernemann, and S. Heckl, *Novel gastrin receptor-directed contrast agents-potential in brain tumor magnetic resonance imaging*, Medicinal Chemistry **8** (2012), no. 2, 133–137.

[108] T. Sugahara, Y. Korogi, M. Kochi, I. Ikushima, Y. Shigematu, T. Hirai, T. Okuda, L. Liang, Y. Ge, Y. Komohara, et al., *Usefulness of diffusion-weighted MRI with echo-planar technique in the evaluation of cellularity in gliomas*, Journal of magnetic resonance imaging **9** (1999), no. 1, 53–60.

[109] V. N. Thijs, D. M. Somford, R. Bammer, W. Robberecht, M. E. Moseley, and G. W. Albers, *Influence of arterial input function on hypoperfusion volumes measured with perfusion-weighted imaging*, Stroke **35** (2004), no. 1, 94–98.

[110] H. S. Thomsen, *Nephrogenic systemic fibrosis: a serious late adverse reaction to gadodiamide*, European radiology **16** (2006), no. 12, 2619–2621.

[111] P. Thullen, J. C. Dudley, D. L. Greene, J. L. Smith, and H. H. Woodson, *An experimental alternator with a superconducting rotating field winding*, Power Apparatus and Systems, IEEE Transactions on (1971), no. 2, 611–619.

[112] D. C. Tong, M. A. Yenari, G. W. Albers, M. O'brien, M. P. Marks, and M. E. Moseley, *Correlation of perfusion-and diffusion-weighted MRI with NIHSS score in acute (< 6.5 hour) ischemic stroke*, Neurology **50** (1998), no. 4, 864–869.

[113] C. Tu and A. Y. Louie, *Nanoformulations for molecular MRI*, Wiley Interdisciplinary Reviews: Nanomedicine and Nanobiotechnology (2012).

[114] F. H. R van der Zande, P. A. M Hofman, and W. H. Backes, *Mapping hypercapnia-induced cerebrovascular reactivity using BOLD MRI*, Neuroradiology **47** (2005), no. 2, 114–120.

[115] Z. Wang, J. M. van Oort, and M. X. Zou, *Development of superconducting magnet for high-field MR systems in China*, Physica C: Superconductivity (2012).

[116] H. Wei, X. Xue, J. Leach, M. J. Padgett, S. M. Barnett, S. Franke-Arnold, E. Yao, and J. Courtial, *Simplified measurement of the orbital angular momentum of single photons*, Optics Communications **223** (2003), no. 1–3, 117 – 122.

[117] E. C. Wiener, S. Konda, A. Shadron, M. Brechbiel, and O. Gansow, *Targeting dendrimer-chelates to tumors and tumor cells expressing the high-affinity folate receptor*, Investigative Radiology **32** (1997), no. 12, 748.

[118] W. Wiltschko and R. Wiltschko, *Magnetic compass of European robins*, Science **176** (1972), no. 4030, 62–64.

[119] G. Woan, *The cambridge handbook of physics formulas*, Cambridge Univ Pr, 2000.

[120] O. Wu, W. J. Koroshetz, L. Üstergaard, F. S. Buonanno, W. A. Copen, R. G. Gonzalez, G. Rordorf, B. R. Rosen, L. H. Schwamm, R. M. Weisskoff, and A. G. Sorensen, *Predicting tissue outcome in acute human cerebral ischemia using combined diffusion-and perfusion-weighted MR imaging*, Stroke **32** (2001), no. 4, 933–942.

[121] K. Yamada, O. Wu, R. G. Gonzalez, D. Bakker, L. Østergaard, W. A. Copen, R. M. Weisskoff, B. R. Rosen, K. Yagi, T. Nishimura, and A. G. Sorensen, *Magnetic resonance perfusion-weighted imaging of acute cerebral infarction*, Stroke **33** (2002), no. 1, 87–94.

[122] H. Zhu, D. G. Buck, Z. Zhang, H. Zhang, P. Wang, V. A. Stenger, M. R. Prince, and Y. Wang, *High temporal and spatial resolution 4D MRA using spiral data sampling and sliding window reconstruction*, Magnetic Resonance in Medicine **52** (2004), no. 1, 14–18.

Index

ADC, 31
Aliasing problem, 80
Alternating symbol, 58
Analytical solution, 5
Angular momenta addition, 68
Angular momentum, 56
Animal models, 31
Anti-commutation relations, 73
Anti-commutator, 73
Antiparallel, 12
Antisymmetric, 75
Apparent diffusion coefficient, 30
Applied magnetic field, 8, 9, 11, 13, 14, 26, 36, 64
Arterial input function, 33
Arterial spin labeling, 33, 37
Associativity, 70
Atom, 62
    atomic orbital, 42
    atomic radius, 42
Atomic physics, 88
Avogadro constant, 27
Axonal
    cytotoxic damage, 33
    edema, 30
    tract, 32
Azimuthal, 44, 45, 52, 54, 55
Azimuthal equation, 44, 45

B_1 perturbation, 16
Band-limited signal, 79
Band-selection, 6
Bandlimit, 80
Bandwidth frequency, 23
Bardeen, 25
BCS theory, 25
Bell state, 85
Bijective homomorphism, 74
Bit, 84
Black-body radiation, 4
Bloch equations, 6, 21
Bloch sphere, 84, 85
Blood perfusion, 33
Bohr
    magneton, 10, 27
    Niels Bohr, 4
    radius, 27, 51
BOLD
    labeling, 33
    MRI, 6, 36
Boltzmann constant, reduced, 16
Bose-Einstein condensation, 25, 88
Boson, 25
    Vector Boson, 87

Bound state, 42
Brain, 8, 19, 31, 32, 35, 37
Brain abscess, 31
Brain territory, 34
Brownian motion, 30
Bulk effect, 9
Bulk magnetization, 9, 12–14

Cardio-cerebral imaging, 39
Carotid vasculature, 34–36, 38
Cerebral artery, 18
Cerebral blood
    flow, 33, 37
    volume, 33, 37
Cerebral ischemia, 33
Cerebrospinal fluid, 4, 8, 30
Cerebrovascular reactivity, 37
Chain rule, 45
Change of basis, 68
Charge, 5, 10
Circle-of-Willis, 36–38
Classical Electromagnetics, 8
Clebsch-Gordan coefficients, 68, 70
Cobalt, 22
CODATA, 52
Coherence, 25
Coherent states, 87
Commutation relation, 63, 72–74
Commute, 42
Compartment model, 33
Computational method, 33
Computed tomography, 11
Computer processor, 24
Condensed matter physics, 88
Convolution theorem, 81
Cooper, 25
Cooper pairs, 25
Copenhagen interpretation, 85
Copper, 22
Coulomb potential, 46
Coupled space basis, 68
Curl operator, 19
Current, 19
    electric, 20
    loops, 25

Damadian, Raymond, 24
Dephase, 17
Dephased oscillations, 14, 15
Dephasing effects, 33
Diamagnetic, 36
Diffusion tensor imaging, 32
Diffusion weighted imaging, 6, 30, 32

INDEX

Diffusion-perfusion mismatch, 33
Dirac, Paul A. M., 4
Discrete Fourier transform, 79
Discreteness property, 42
Disease, 26
Double cover, 72
DTFT, 80
Duality of function, 25
DWI, 30, 33
Dynamic susceptibility, 33

Earth's magnetic field, 25
Eigendirection, 32
Eigenfunction, 42, 56
Eigenstate, 42, 68
Eigenvalue, 42, 62, 64
    problem, 47
    problem solver, 42
Einstein summation, 74
Einstein, Albert, 85
Electric
    constant, 27
    field, 19
Electrical
    neuroactivity, 37
    resistance, 25
Electromagnetic
    induction, 19
    signals, 11
Electromotive force, 19, 20
Electron, 6, 8–11, 13, 26, 42, 46, 47, 62, 68, 72, 86
    cloud, 42
    mass, 10
    pairing, 25
    volt, 27
Electronic orbital configuration, 42, 46
Electronics, 88
Elementary charge, 10
Elementary particles, 5, 62
Energy
    difference, 12
    eigenstate, 13
    eigenvalue, 13, 51
    gap, 26
    penalty, 26
    state, 12
Ensemble, 8, 26
EPR paradox, 86
Equilibrium magnetization, *see* Magnetization
External carotids, 37

Faraday-Maxwell equation, 8, 19, 21–23, 25
Fast algorithms, 78
Fast Fourier transform, 78
Fast multipole method, 78
Fast spin echo, 17
Fast spoiled gradient echo, 34, 38
FeO, 22
Fermi exclusion principle, 25
Fermion, 25
FID, 6, 20
    envelope, 33
    signal, 25
FLAIR, 8, 18
    T_2, 18
Fock state, 87
Force, 6, 11, 13
Fourier transform, 78, 81
Free inductance decay, 6, 78
Free inductance signal, 19
Frequency domain, 78
Functional MRI, 37

g-factor
    electron, 27
    proton, 27
Gadodiamide, 22
Gadolinium, 33
    Gadolinium ion, 22
    Gadolinium-DPTA, 22, 40
Gell-Mann matrices, 74
Generalized Laguerre equation, 48, 50
Generalized linear model, 34
Glioma, 32
Gluon, 8, 72
Glycine, 35
GPU CUDA, 78
Gradient coil, 20, 25
    X coil, 21
    Y coil, 22
    Z coil, 23
Graphical User Interface, 24
Graphics, 78
Green's function, 20, 33
Ground state, 52, 68
Group generators, 74
Group theory, 5, 72
Gyromagnetic factor, 10
Gyromagnetic ratio, 5, 10, 42
    electron, 27
    proton, 27

Hamiltonian, 9, 13, 42, 43
Hardware, 87
Hartree-Fock, 42
Heisenberg
    Werner Heisenberg, 4
    uncertainty principle, 42
Hematocrit, 37
Hemoglobin, 36
Hermitian, 42
Hertz, 10
Hierarchy of scales, 6, 8
High energy particle physics, 72
High performance computing, 78
Hilbert, David, 4
Hydrogen, 42
    atom, 5, 42, 46, 53, 68
    fine structure, 14
Hypercapnia, 37
Hypothermia, 33
Hypoxic-ischemic encephalopathy, 35

Image reconstruction, 79–81
Imaging, 78
Impulse response, 33
Infinitesimal generators, 72
Integral transformation, 19
Intensive care unit, 36
Interdisciplinary, 25
Intravenous contrast, 6, 21
Intrinsic spin, 5, 9, 42, 62
Inverse Fourier transform, 79
Inversion recovery, 18
Iron oxide, 17
Ischemic penumbra, 6, 33
Isomorphism, 48, 62, 72, 74
Iterative, 23, 42

INDEX

Josephson junctions, 87

Kronecker delta, 73

Lactate, 34
Ladder operator, 58, 69
Laguerre
    associated functions, 50, 53
    associated polynomials, 48, 51
Laplace equation, 43, 46, 48, 56
Larmor
    frequency, 9, 13, 23
    precession, 6, 11
Laser science, 88
Lauterbur, Paul, 24
Legendre differential equation, 45
Levi-Civita symbol, 58
Lie algebra, 62, 72, 73
Liquid helium, 25, 26
Liquid nitrogen, 25, 26
Localization
    3D, 23, 25
Low pass filter, 80
Luttinger liquid, 25

Maghemite, 22
Magnet, 9, 24
Magnetic dipole, 10
Magnetic dipole moment, 36
Magnetic field, 19
    gradient, 23
    inhomogeneities, 33
    pulsed, 25
    static, 25
Magnetic moment, 5, 8, 11, 13
    electron, 27
    proton, 27
Magnetic potential energy, 6, 11, 12
Magnetic properties, 8
Magnetic resonance, 72
    angiography, 6
    spectroscopy, 6, 34
Magnetic spin eigenstate, 13
Magnetism, 6, 9
Magnetization, 6, 12
    equilibrium, 6, 12–14, 17
    longitudinal, 12
    transverse, 17
    vector, 19, 26
Manganese, 22
Mass
    electron, 27
    neutron, 27
    proton, 27
    reduced, 47
Matrix-matrix multiplication, 72
MCA, 18, 30–32, 39, 40
Mean transit time, 33
Measurement, 69
Mebrahtu, Henok, 25
Model, 8, 12
Molar gas constant, 27
MRA, 6, 34–38
    4D MRA, 39
MRI, 5, 8, 11, 25, 26, 30, 42, 52, 78, 79
    machine, 24
    machine components, 6
    probe, 20
Multi-core processors, 78
Multiple sclerosis, 18, 31

N-acetyl aspartate, 34
Nanoparticle, 22
Nanotesla, 25
Natural logarithm, 45
Near-infrared spectroscopy, 37
Neonate, 35
Nephrogenic systemic fibrosis, 22
Neuronal pathways, 32
Neutron, 9, 26, 62, 68
Nickel, 22
NMR, 87
Noise filtering, 78
Non-ionizing radiation, 11
Non-relativistic, 43
Nuclear magnetic resonance, 87
Nuclear magneton, 10, 27
Nuclear spin, 9, 68
Nucleon, 6, 8–11, 13, 72
Nucleus, 5, 42, 58, 68, 72
Numerical computation, 20
Nyquist
    Nyquist criterion, 80
    Nyquist frequency, 79

Observable, 42
Occipital, 40
Operator, 64
    normal bounded, 42
    splitting, 47
Optical lattices, 87
Optical technology, 8
Optical tweezers, 88
Orbit, 42
Orbital angular momentum, 5, 9, 26, 42, 43, 56, 63, 68, 72, 87
    addition, 5
    of light, 88
    operator, 56
Ordinary differential equations, 33, 44
Orthogonal matrix, 72
Orthogonalize, 69
Oxygen, 36
    consumption rate, 37

Parallel, 12
Paramagnetic, 36
Parametric down-conversion, 88
Paraventricular, see Periventricular
Parieto-Occipital, 31
Pauli
    Pauli, Wolfgang, 62
    spin matrices, 73
PCA, 18, 31, 32, 40
Pedagogically separable, 9
Penumbra, 6, 34, 37
Perfusion model, 33
Perfusion weighted imaging, 6, 33
Periodic summation, 80
Periventricular, 6
    white-matter, 18
Permanent magnet, 25
Perturbation, 26
Phase, 9
Phase contrast MRA, 38
Phased oscillations, 14
Phased precession, 15
Phonon exchange, 25
Phosphorus, 35
Photon, 87
Photonics, 88
Planck

Max Planck, 4, 87
Planck constant, 27
Poincaré, Henri, 72
Poisson summation formula, 80
Polar, 44
    polar equation, 44, 45
Polarization, 87
Posterior circulation, 17, 31
Potential Energy, 6
Precession, 9, 11
Principal component analysis, 32
Probability, 4, 58, 85
Probability amplitude, 84, 85
Probability density function, 42
Proton, 5, 9, 10, 12, 46, 47, 68
    density, 16
    mass, *see* Mass
Pulse, 11
    pulse duration, 14
    pulse excitation, 15, 17, 19, 23
    pulse magnetization, 6, 9
    pulse sequence, 25, 78
PWI, 33, 37

Quantum chromodynamics, 72
Quantum communication, 86, 87
Quantum computation, 87
Quantum computing, 86
Quantum cryptography, 87
Quantum cyber security, 87
Quantum dots, 87
Quantum electrodynamics, 10
Quantum entanglement, 85, 86
Quantum field theory, 8
Quantum information theory, 84, 85
Quantum mechanics, 8, 9, 26, 42
    postulate, 43
Quantum number, 13, 58
    azimuthal, 52, 54, 55
    magnetic, 54, 55
    of rotation, 55
    principal, 51, 54
    radial, 51, 52
Quantum optics, 87
Quantum phase transition, 25
Quantum superposition, 86
Quantum tunnelling, 25
Quark, 5, 8, 62, 72
Qubit, 84–87

Radial, 44
Radial equation, 52
Radial solutions, 53
Radial wave function, 42, 55, 58
Radio frequency, 11
    RF coil, 19, 20, 24
    RF waves, 8, 9
    RF-canceled, 33
Randomized spin orientations, 9
Reality property, 42
Receptors
    tissue-specific, 22
    tumor-specific, 22
Red blood cell, 36
Regularity, 44
Regularity condition, 48
Representation, 68, 69, 72, 79
Resonance frequency, 23
RF, *see* Radio frequency
Rydberg
    Rydberg constant, 52

Rydberg energy, 27

Sampling, 79
    non-uniform sampling, 78
    sampling rate, 80
Schrödinger
    eigenvalue problem, 42
    Erwin Schrödinger, 4
    Schrödinger equation, 5, 9, 43, 46
    time-independent, 53
Schrieffer, 25
Sensorimotor, 37
Separation of variables, 43, 47
Set theory, 34
Shannon sampling theorem, 79
Signal processing, 5, 25, 78
Signal-to-noise, 78
Sinc function, 80
Singlet state, 69
Slice selection, 23
Slice thickness, 23
$SO(3)$, 5, 62, 72
$SO(N)$, 72
Software instruction, 25
Solid state physics, 88
Space localization, 6
Spectral theorem, 42
Speed of Light
    in vacuum, 27
Sphere, 56
Spherical coordinates, 42, 43, 46, 56
Spherical harmonics, 42, 46–57
Spin, 5, 10, 26, 62, 68, 69, 72, 87
    spin 1, 87
    spin 1/2, 13, 72, 86
    spin distribution, 12
    spin observable, 72
    spin saturation, 38
Spin configuration, 68
Spin geometry, 63, 64
Spin-Lattice, 6
Spin-Lattice relaxation, 16
Spin-Orbit coupling, 5
Spin-spin, 6
State transition, 9
Statistical mechanics, 8
Stokes theorem, 19
Stroke, 6
    acute ischemic, 17, 30–32, 34, 37, 40
    hemorrhagic, 31
    occipital, 4
    quantitation scales, 33
Strong force, 72
Strong interaction, 72
Structure constants, 75
Sturm-Liouville problem, 45
$SU(2)$, 5, 62, 72, 73
$SU(3)$, 74, 75
$SU(N)$, 72
Subatomic, 4, 8
Subatomic particle, 26
Subcortical infarction, 31
Subnuclear, 8
Superconducting electromagnet, 25
Superconductivity, 25
Superimposition, 85
Superparamagnetic iron oxides, 23
Superposition, *see* Quantum superposition
Susceptibility effects, 33

INDEX

Susceptibility weighted imaging, 39
SWI, 39, 40
Symmetry, 72, 75

T_1, 8
T_1 recovery, 6, 16
T_2
    decay, 6
    MRI, 4
    relaxation, 17
    signal, 33
T_2^*, 33
Technology, 15, 26
Temperature, 12
    absolute, 12
    subcritical, 26
Tesla, 10, 25
Thrombolytic therapy, 33
Time domain, 78
Time evolution, 9, 21
Time to peak, 33
Time-of-flight
    2D, 34, 38
    3D, 35, 38
    MRA, 38
Tissue, 26
Torque, 6, 8, 11, 12
Tractography, 32
Transient cerebral ischemia, 31
Transverse phase, 14
Traumatic brain injury, 31
Triplet state, 69

Ultraviolet catastrophe, 4
Uncoupled basis, 69, 85
Unit determinant, 72
Unitary matrix, 72
Unitary transformation, 85
Unpaired electron, 10

Vitreous, 4, 8, 30
von Neumann, John, 4
Voxel, 32

Wave function collapse, 85
Whittaker-Shannon interpolation, 79
Wire coil, 25

X-ray, 11

Zeeman effect, 14

Made in the USA
Lexington, KY
05 November 2012